LIFE, TEMPERATURE, AND THE EARTH

:: LIFE, TEMPERATURE, AND THE EARTH

The Self-organizing Biosphere

David Schwartzman

COLUMBIA UNIVERSITY PRESS   NEW YORK

Columbia University Press

Publishers Since 1893

New York   Chichester, West Sussex

Copyright © 1999 Columbia University Press

Library of Congress Cataloging-in-Publication Data

Schwartzman, David (David W.)

    Life, temperature, and the earth : the self-organizing biosphere / David Schwartzman.

      p.   cm.—

    Includes bibliographical references.

    ISBN 978-0-231-10212-4 (cloth); ISBN 978-0-231-10213-1 (paper)

    1. Biosphere.   2. Bioclimatology.   3. Weathering.   I.Title.

II.Series.

QH343.4.S36   1999

577.2′2—dc21   99-25856

To my sons, Sam and Peter; my father, Max; and the four who inspired whatever value is contained in this book: Jim Lovelock, Lynn Margulis, Vladimir Vernadsky, and Fred Engels

CONTENTS

Acknowledgments   ix

Introduction: A Personal Note   xi

1: Climatic Evolution: From Homeostatic Gaia to Geophysiology   1

2: The Biogeochemical Cycle of Carbon   15

3: Faint Young Sun Paradox and Climate Stabilization   32

4: Weathering and Its Biotic Enhancement   43

5: Weathering: From Theory and Experiment to the Field   66

6: Quantifying the Biotic Enhancement of Weathering and Its Implications   80

7: Surface Temperature History of the Earth   99

8: Did Surface Temperatures Constrain Microbial Evolution?   119

9: Self-organization of the Biosphere   157

10: Alien Biospheres?   179

11: Conclusions   191

References   197

Index   239

:: ACKNOWLEDGMENTS

The author acknowledges the close collaboration of Tyler Volk, Mark McMenamin, Mike Rampino, Ken Caldeira, and Steve Shore, along with the help and patient advice of Connie Barlow, Scott Bailey, Bob Berner, Susan Brantley, Ford Cochran, Paula DePriest, Dave Des Marais, Tim Drever, Sam Epstein, Jack Farmer, Bruno Giletti, Peter Gogarten, John Grotzinger, David Hawksworth, Marti Hoffert, Dick Holland, Togwell Jackson, Annika Johansson, Jim Kasting, Lee Klinger, Paul Knauth, Lee Kump, Jim Lawrey, Franz May, Euan Nisbet, Verne Oberbeck, Greg Retallack, Norrie Robbins, Mike Russell, Peter Schultz, Paul Shand, Rod Swenson, Bill Ullman, Peter Westbroek, Art White, and my colleagues at Howard University.

At the age of 12 or 13, I read a J. B. S. Haldane book on the natural sciences, which I discovered in my uncle's library. Haldane mentioned some of the Russian work on the biological role in weathering (probably Polynov or one of his students). I had forgotten about all this until I became interested in this subject again as an adult. I have not yet found the Haldane volume mentioned, but those early researchers felt, doubtlessly influenced by Vernadsky's thinking on the subject, that weathering was basically a biological phenomenon and would be much slower without life present. Having grown up in an "old left" household in the 1950s in Brooklyn, I discovered the Marxist classics in a hidden space below the family television. Reading Engels's *Dialectics of Nature* strongly impressed me. Burning the top of my dresser with chemistry experiments and collecting minerals, insects, and plants occupied my childhood. I probably passed Stephen Jay Gould on my monthly pilgrimage to the mineral hall at the Museum of Natural History (the dinosaur exhibit was on the way). I majored in chemistry at Stuyvesant High School and geochemistry at City College of New York and Brown University, where my graduate research was on excess argon in the Stillwater Complex and degassing models of the Earth. I did a term paper on Vernadsky and biogeochemistry as a senior in Alexander Klots's (the author of *A Field Guide to Butterflies*) biology class in the spring of 1964.

This book is an outgrowth of research that I have been pursuing for the past 15 years, since I first felt the powerful heuristic influence of Lovelock's 1979 book on Gaia. The concept of Gaia strongly resonated with my sense that spheres of nature interacted dialectically in the Engelsian sense, that is, emergent phenomena arise from the interactions of the parts (the whole's systems and subsystems; for a lucid exposition of a modern dialectics of nature, see Levins and Lewontin 1985). The whole, Gaia, evolves as the parts (organisms and ecosystems) themselves evolve. The Gaian interactions to be discussed in this book include those among life, climate, weathering, hydrology, and crustal/impact history. This book will present in a systematic way the developing theory for a biotically mediated regulation of Earth's temperature over geologic time, the first order determination of the history of climate. The emphasis is on long-term geologic trends, not the short-term perturbations that have received so much media attention (e.g., the anthropogenic greenhouse effect).

The first third of the book (chapters 1–3) will introduce my theory of biospheric evolution followed by the Gaia concept and its evolution in the 1980s and 1990s. The biogeochemical cycle of carbon and the silicate-carbonate climatic stabilizer will then be discussed. This stabilizer entails the dependence of the chemical weathering rate (the sink for atmospheric carbon dioxide) on global temperature. A key question raised here is the criticality of life to the operation of the stabilizer. Is climatic stabilization an emergent property of the Earth's biosphere, in the context of changing solar luminosity and other abiotic factors?

The second third (chapters 4–6) will present a systematic exposition of the weathering process, including recent research on its biotic enhancement, and a model for understanding the habitability of the Earth over geologic time. The evidence for biotic enhancement of weathering includes experimental and field studies that need to be significantly expanded. This section will include discussion of the abiotic factors affecting climatic evolution, such as tectonics and the carbon geodynamic cycle, as well as their possible biotic mediation.

The last third of the book (chapters 7–11) will present a reinterpretation of the surface temperature history of the Earth. A much warmer Precambrian Earth surface is supported by a diverse body of evidence, implying a high present biotic enhancement of weathering consistent with our earlier estimates (two orders of magnitude, though probably not three). A geophysiological theory of the coevolution of life and the biosphere itself, entailing a progressively changing biotic enhancement of weathering, will be presented. The implications of these results to evolutionary biology and to bioastronomy (the search for life elsewhere in the universe) and a theory of the self-organization of the Earth's biosphere will be discussed. One startling conclusion emerges: microbial evolution in the Precambrian was constrained by the surface temperature, indicating that the major events in biotic evolution were forced by the physical context for self-organization. A concluding chapter will summarize the main conclusions and raise a research agenda for diverse fields of science ranging from geochemistry to biology.

1 CLIMATIC EVOLUTION: FROM HOMEOSTATIC GAIA TO GEOPHYSIOLOGY

A theory of biospheric evolution. Coevolution of climate and life (precursors: co-evolution as life adapting to changing climate; Preston Cloud: oxygen and life; Schneider and Londer: anthropogenic effects on climate) A brief history of the Gaia concept: Lovelock's early papers, Doolittle's challenge, DaisyWorld, the AGU Gaia Conference, 1988. A Gaian mechanism on today's Earth? (DMS and cloud formation over the ocean, a climatic stabilizer?)

A Theory of Biospheric Evolution

I begin with a brief outline of my theory of how our biosphere has evolved since the origin of life some 4 billion years ago. This theory is a provocation: to think about the evolution of life and climate in a new way. Both the beginning and end of life on this planet are determined by purely nonbiological conditions, the beginning by the hydrothermal activity on the ocean floor, where the origin of life took place, the end by the rising radiant energy flux from the sun. But between these times, predetermined by the initial conditions of our solar system, the biosphere evolves in its overall patterns deterministically, going from a hothouse with surface temperatures near 100°C to an icehouse, with intermittent glacial periods, then in the future back into a hothouse regime before its destruction. A progressive increase in the diversity of habitats for life and a concomitant biotic evolutionary explosion has occurred in the past 4 billion years, only to be reversed in the future with a return to the hothouse.

The surface temperature scenario argued for in this book is shown in figure 1-1. Now for a key concept in this theory of biospheric evolution. Sur-

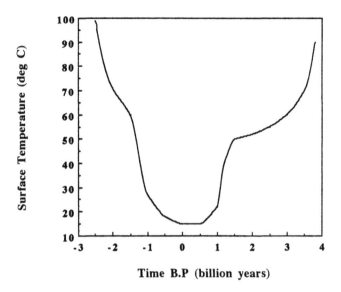

FIGURE I-I.
Surface temperature history of Earth as argued for in this book. Surface temperatures
(°C) versus time BP (billion years). Negative time corresponds to the future.

face temperature is a critical constraint on the tempo of major events in biotic
evolution, while it is determined itself by a progressively increasing role of
biota in climatic change over geologic time, within the context of abiotic
evolution (solar and terrestrial). The temperature constraint has occurred because each major innovation in biological evolution, such as oxygenic photosynthesis (emergence of cyanobacteria), has an inherent biochemical and
biophysical upper temperature limit for its metabolism. Thus, with the long-
term cooling of the Earth's surface, new metabolisms and cell types became
possible as their upper temperature limit was reached. Cooling occurred because of the combined effects of abiotic variations, such as volcanic outgassing rates and rising solar luminosity, and the progressively powerful effect of
land biota on the sequestering of carbon from the atmosphere by the chemical weathering process in soils. As a result, carbon dioxide levels in the atmosphere have dramatically declined since the origin of life, declined enough
to have decreased surface temperatures from near 100°C to the present
global mean of 15°C, despite the rising solar energy flux. Where did this
carbon dioxide go? Some was probably recirculated down into the mantle,

but the crust now contains the equivalent of some 60 times the total pressure of the atmosphere, in the form of limestone and marble (calcium carbonate), which was sequestered from atmosphere.

The biosphere has evolved deterministically as a self-organized system, given the initial conditions of the sun-Earth system. The origin of life and the overall patterns of biotic evolution were highly probable outcomes of this deterministic process. These overall patterns include the emergence of oxygenic photosynthesis and the history of endosymbiogenesis, which resulted in the emergence of Eukarya (complex life) and its kingdoms. Evolution of procaryotes and complex life on terrestrial planets around sunlike stars are expected to have similar geochemical and climatic consequences. Thus, the main patterns would be conserved if "the tape were played twice," a theory argued from computer simulations by Fontana and Buss (1994). The width of the habitable zone for Earth-like planets around sunlike stars for complex life may be substantially smaller than that for the appearance of biota, constrained by the presence of liquid water. Surface temperature history on terrestrial planets may be critical to the time needed to evolve complex life and intelligence. Biotically mediated cooling increases the width of the habitable zone for the possible occurrence and evolutionary time frame of complex life. For Earth-like planets within the habitable zone of stars less massive than the sun, the earlier emergence of complex life is expected, all other factors being the same.

This is a brief summary of my theory, which grew out of collaborative research, first of all with Tyler Volk, and the input of a vast literature. The rest of this book will explore step by step the science behind this theory, starting in the next chapter with the biogeochemical cycle of carbon, which determines at any time the level of carbon dioxide in the atmosphere and the surface temperature. But first a historical overview of theories of the coevolution of life and its environment (the most common definition of the biosphere) is needed. The Gaia concept and its development figure prominently in this history. We begin with Gaia for two reasons: first, its historical importance in the development of coevolutionary theory; and second, the Gaia theory has had a profound heuristic influence on a global network of scientists from many disciplines, including myself (Schneider and Boston 1991). Gaia theory has stimulated an expanding wave of research into the coupling of life and its environment, paradoxically despite and because of its impurity and metaphorical excursions.

Coevolution of Climate and Life

Although all researchers accepting the scientific paradigm agree that life has indeed evolved over geologic time, there is still debate and some uncertainty as to whether climate did also, particularly in the strong sense of some directional vector, a mode that is certainly debatable with respect to biological evolution. The temperature record of the past billion years supports fluctuations of some 5 to 10°C from the present mean global surface value of 15°C, but the interpretation of the more ancient record, back to 4 billion years ago (4 Gigayears [Ga]) is more confused, with some researchers supporting a nearly constant temperature, and others a strong cooling to present. However, one aspect of climate has almost certainly changed over the past 4 billion years—atmospheric composition, particularly the oxygen level.

The geologic/geochemical record generally supports low to exceedingly low atmospheric pO_2 levels prior to about 2.2 Ga, with modern levels of some 0.2 bar being approached in the Phanerozoic. This evidence includes the presence of minerals deposited on the surface prior to about 1.8 Ga that are unstable in free oxygen, abundant "red beds" later than this date, and inferences based on the variation in oxidation state of iron in ancient soils (paleosols) of different ages (Holland 1994).

This evidence deserves closer attention. First, detrital (grains deposited by surface water) uraninite (UO_2) is found in large deposits with ages older than 2.3 Ga. Uraninite quickly oxidizes in the presence of free oxygen. Abundant red beds (sediments with iron in the oxidized ferric state) only appear in the geologic record by 2.3 to 2.4 Ga. Finally, paleosols from before 2.2 Ga show apparent primary leaching of iron, indicating low atmospheric oxygen levels because ferrous, not ferric, iron is easily dissolved in ground water. Paleosols from after 1.8 Ga have oxidized iron. Based on this evidence, Holland (1994) postulated that atmospheric oxygen rose significantly from 2.2 to 1.9 Ga. From their study of the kinetics of calcium carbonate precipitation and its influence on the textures of carbonate rocks of Precambrian age, Sumner and Grotzinger (1996) concluded that high concentrations of ferrous iron were present in the Archean ocean, only possible with low oxygen levels, with ferrous iron levels declining and oxygen rising in the atmosphere at 2.2 to 1.9 Ga. This scenario is based on their observation that ferrous iron inhibits calcite precipitation, which would result in microcrystalline textures, while apparently allowing the precipitation of the fibrous herringbone cal-cite found abundantly in Archean carbonates.

There are, however, dissenting voices to this scenario. Towe (1994) has long argued for the presence of modest levels of oxygen in the Archean atmosphere (3.8 to 2.5 Ga) in contrast to the more commonly held view of geochemists and paleoclimatologists that levels then were exceedingly low (Kasting 1987). An even more radical challenge has come from Ohmoto (1996, 1997a, 1997b) who has looked closely at the paleosol record, con-cluding that both oxidized and reduced primary iron occurs in the record both before and after 2.2 Ga. He argues for a comparable oxygen level in the 2.2 to 3.0 Ga atmosphere to the present atmospheric level (PAL). Because the other body of evidence seems to support Holland's scenario (Holland and Rye 1997), it is not clear at this point how reliable the paleosol evidence really is, given the possibilities for alteration of the original weathering im-print in the past 2 billion years or more.

On the basis of an inferred variation in atmospheric pO_2 levels, Preston Cloud (1976) argued for atmospheric oxygen being a constraint on biotic evolution, with the emergence of first eucaryotes and then Metazoa linked to progressive increases in the levels of atmospheric oxygen. Cloud argued that at least 1% PAL free oxygen was needed for eucaryote (mitochondrial) metabolism, still higher levels for megascopic algal and metazoan forms that exchange gases by diffusion. Finally, larger skeletonized Metazoa and land plants require near modern levels. Of course, a feedback from life to atmo-spheric composition is also required, the generation of oxygen by photo-synthesis coupled with burial of organic carbon. Others have highlighted climate/life coevolution, but restricting this linkage to recent times (e.g., Schneider and Londer 1984, pointing out Pleistocene glacial/interglacial cycles and anthropogenic effects on climate).

Gaia

A much more radical conception of coevolution was put forward by James Lovelock, an atmospheric chemist, soon joined by Lynn Margulis (the biolo-gist best known for her theory of endosymbiogenesis), namely the Gaia hy-pothesis. Gaia is not a rephrasing of "coevolution": "Coevolution is rather like a platonic friendship. The biologist and the geologist remain friends but never move on to an intimate, closely coupled relationship. Coevolution theory includes no active regulation of the chemical composition and climate of the Earth by the system comprising the biota and their material environment" (Lovelock 1989). What Lovelock is arguing here is that the classical view of coevolution of climate and life does not capture the richness of interactive processes and feedbacks, nor does it recognize that planetary biota actively determines its planetary environment. But Lovelock goes even further asserting that in some sense the Earth (surface) is living, with its own physiology, a geophysiology of a superorganism. Homeostasis is a characteristic of animal metabolism. Biotic regulation of its global external environment leading to, for example, stable climate is for Lovelock homeostasis on a planetary scale.

One of the most famous Gaian metaphors is the "living Earth" (just how metaphorical or literal depends on the text one reads). The Earth as a superorganism resonates with Hutton's conception. This 18th-century Scottish doctor, farmer, and arguably the founder of modern geology, did his thesis entitled "The Blood and Circulation in the Microcosm," drawing back from the medieval conception of the macrocosm and microcosm (Adams 1954). Hutton wrote, "We are . . . thus led to see a circulation in the matter of the globe, and a system of beautiful economy in the works of nature. This earth, like the body of an animal, is wasted at the same time that it is repaired" (quoted in McIntyre 1963). Hutton's equivalent of blood in the circulation of the Earth was water.

Much has been written about the origin of the Gaia hypothesis. Lovelock's own account is the most eloquent (Lovelock 1979). Briefly, Lovelock, working at the Jet Propulsion Laboratory in the 1960s, concluded that the atmospheric composition of Mars should be indicative of the presence or absence of life. Several constituents in the Earth's atmosphere, particularly oxygen and methane, are not in equilibrium with its crust, with measured fluxes being dramatically different from those expected on an abiotic Earth (figure 1-2). Thus, he concluded that any disequilibrium of Mars' atmosphere with its crust should be strong evidence for life (as it turned out, Mars' atmosphere is apparently close to chemical equilibrium with its crust, consistent with the present absence of a living surface biota, and the hegemonic consensus from the results of the Viking biology experiments; Gilbert Levin, the principal investigator for the labeled release experiment, has been one continuing dissenter).

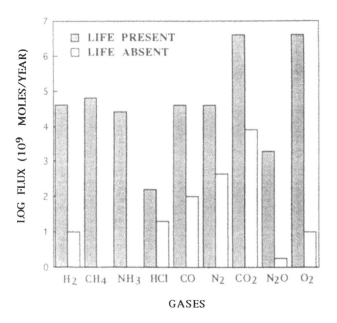

FIGURE 1-2.
The fluxes of gases to the present atmosphere compared with those expected for an abiotic Earth (after Lovelock 1989).

Lovelock's insight led to a radically new explanation of Earth's habitability for the past 3 billion years (now accepted to be at least 3.5 billion years based on fossil evidence). This habitability was not just "dumb luck," but rather a result of continuous biotic interaction with the other components of the biosphere, the atmosphere, ocean, and soil/upper crust. The requirements of habitability include favorable temperatures, ocean salinity, and—at least for the past 2 billion years—atmospheric oxygen levels for aerobes. In Lovelock and Margulis's early papers, we find a formulation of Gaia as a homeostatic system:

From the fossil record it can be deduced that stable optimal conditions for the biosphere have prevailed for thousands of millions of years. We believe that these properties of the terrestrial atmosphere are best interpreted as evidence of homeostasis on a planetary scale maintained by life on the surface (Lovelock and Margulis 1974a).

The notion of the biosphere as an active adaptive control system able to maintain the earth in homeostasis we are calling the Gaia Hypothesis (Lovelock and Margulis 1974b).

What are optimal conditions? Optimal for maximum productivity of ecosystems, the global biota? Optimal for the persistence of planetary biota, but which components? Is optimality to be measured in number of species? (If so, on the present Earth beetles apparently win out.) Did the anaerobic procaryotes of the Archean optimize atmospheric conditions for their successors, the aerobes? Optimality is a problematic concept at the very least.

After the publication of Lovelock's first book (1979), homeostatic Gaia came under heavy attack in the 1980s primarily from staunch neodarwinian biologists (Doolittle 1981, Dawkins 1982, Maynard Smith 1988). They objected to the concept of life optimizing its external conditions by natural selection because the biosphere is a single entity "competing" against no other (see discussion in Barlow and Volk 1992a). Furthermore, "Gaia, as a cybernetic system, must have mechanisms for sensing when global physical and chemical parameters deviate from optimum, and mechanisms for initiating compensatory processes which will return those parameters to acceptable values (negative feedback)" (Doolittle 1981).

In response to such criticism, Watson and Lovelock (1983) developed the Daisyworld model, an attempt to demonstrate the possibility of planetary surface homeostasis without invoking natural selection. This model in its simplest form assumes dark and light daisies populating a planet, subject to a steadily rising energy flux from outside (as the Earth/sun couple). The daisies differ only in their reflectivity (albedo) of incoming radiation, with the same growth and death rates. The result is that the planetary temperature significantly stabilizes as the incoming energy flux increases, an outcome of the "thermostat" set up by the successive expansion of first dark then light daisies, each affecting the planetary albedo (figure 1-3). Lovelock (1989) and others (Saunders 1994) have followed up this original model by making the "ecosystem" more complex (e.g., adding predators, shades of daisies) and more realistic controls on heat flow between regions on the hypothetical planet Daisyworld. The model results appear to strengthen the hypothesis that homeostasis, at least by albedo modification, would result from plausible biotic physiology, without natural selection operating to guarantee optimization.

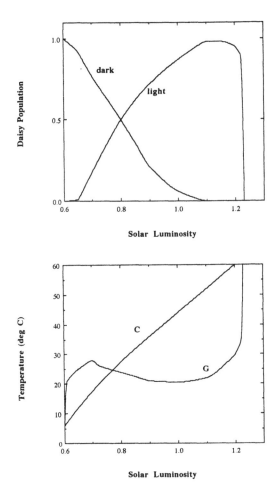

FIGURE I-3.

Models of evolution of Daisyworld. The top graph shows the variation of dark and light daisy populations in arbitrary units as a function of a variable solar luminosity (1 = present sun). The bottom illustrates the temperature history according to a conventional assumption (C) that life (daisies here) only adapts to temperature changes, without affecting the surface temperature of the planet, and the Gaian Daisyworld model (G) where the differential growth response of light and dark daisies results in regulation of global temperature by changing the surface albedo (after Lovelock 1989).

Lenton (1998) reviewed more recent work, including his own research, which demonstrated a robust self-regulation in Daisyworld with mutation and natural selection. This evidence supports the continued fruitfulness of looking for self-regulation in ecosystems and the biosphere itself. However, under conditions of interspecies competition, surface temperature in Daisyworld can change more widely than without the presence of daisies (Cohen and Rich 1998).

The main use of Daisyworld to the Gaia hypothesis is its rebuttal to neo-darwinian criticism. Curiously, Margulis and Lovelock (1974) suggested possible albedo-linked regulation of the Earth's climate, at least for the Precambrian, one that did invoke natural selection. In this model, oceanic algae stabilized surface temperature by their mutation–induced variable albedo, optimizing local temperature, assuming their areal extent would be sufficient, the global. Here, external conditions did not merely favor one biotic variety, but created opportunity for its emergence by mutation and natural selection. In this scenario, biotic evolution is linked to climatic stabilization.

However, the Daisyworld model was attacked as invoking arbitrary parameters that enhanced homeostasis, with alternative assumptions leading to destabilization. For example, if black daisies have a higher optimal temperature than white daisies, rather than the same as assumed by Watson and Lovelock, two equilibrium temperatures for Daisyworld could result, with black and white daisies dominating the environment for each value of solar luminosity, temperatures bouncing back and forth (see Kirschner 1989). Homeostatic Gaia was also questioned from another ground, the actual geological record, which indicates something like an oxygen catastrophe brought about by existing anaerobic biota some 2 billion years ago. This event could hardly be viewed as an optimization of conditions for existing biota, at least those living at the surface, subject to poisonous oxygen.

In 1988, the first scientific conference on the Gaia hypothesis took place, chaired by S. Schneider and P. Boston (see Schneider and Boston 1991 for the expanded proceedings of this meeting). This diverse meeting with many critics (most fortunately open-minded) apparently provoked Lovelock to reconsider his original hypothesis. The challenge to homeostatic Gaia was met by Lovelock in his reformulation as geophysiological Gaia. Restated, Gaia is now "a theory that views the evolution of the biota and of their material environment as a single, tightly coupled process, with the self-regulation of climate and chemistry as an emergent property" (Lovelock 1989). Thus, the biosphere is now seen as an evolving system with negative feedback such as climatic stabilization.

However, homeostatic Gaia survives in the more recent writings of both Margulis and Lovelock. As is the case of many parents, they are extremely reluctant to let their progeny go off on its own. For a recent example, note the following:

> The steadiness of mean planetary temperature for the past three thousand million years, the 700 million year maintenance of Earth's reactive atmosphere . . . point to mammal-like purposefulness in the organization of life as a whole . . . Planetary physiology . . . is the holarchic outcome of ordinary living beings. The "purposefulness" of Gaian self-maintenance derives from the living behavior of myriad organisms (Margulis and Sagan 1995, pp. 47–48).

The fundamental challenge to adherents of homeostatic Gaia is for them to demonstrate the fact and mechanisms of self-regulation of the biosphere by and for the biota's benefit, with the functioning of the biosphere as a truly cybernetic system. Geophysiological Gaia does not necessarily require regulation by and for the existing biota, nor optimization in any sense, only that self-regulatory mechanisms emerged during the lifespan of the biosphere. As the original Gaia hypothesis, homeostatic Gaia continues to provoke fruitful research and interesting proposals with possible heuristic value (e.g., Markos 1995; Williams 1996; see discussion in chapter 10; also see Kump's 1996 account of Gaia in Oxford II meeting, April 1996). Lenton (1998) raised the fundamental question: how can self-regulation at all levels of the biosphere emerge from natural selection at the individual level? One key challenge is to identify sensing mechanisms evolved by individual organisms, which in turn allow optimization of whole ecosystems, extending to the global biosphere. However, a number of cautions should be kept in mind in this pursuit. First, stability of conditions does not necessarily entail homeostasis. Near steady states can be achieved without any sensors or cybernetic control machinery; for example, long-term atmospheric moisture levels from the balancing of evaporation and precipitation (Williams 1992). Another example could be salinity levels in the ocean, a steady state being achieved from the long-term equality of incoming dissolved salts and precipitation in tidal flats.

Second, biological regulation may well be limited to restricted "phase space"—the matrix of physicochemical variables—in biospheric evolution, affecting some conditions but not others (e.g., ocean pH but not salinity) without constituting a global homeostatic system. Another possibility is that for global or regional ecosystems homeostasis alternates with periods of drift for a given regulated parameter in phase space. This mode has been called intermittent Gaia. Alternatively, homeostatic regulation in some habitats may have persisted since the origin of life. Could this be the case for the "deep hot biosphere" of the subsurface? If such mini-Gaias do or did exist, it could have important implications to understanding biospheric evolution.

A closely related possibility is "homeorrhetic" Gaia, homeorrhesis denoting shifting steady states. As Lovelock (1991) put it: "Gaia's history is . . . characterized by homeorrhesis with periods of constancy punctuated by shifts to new, different states of constancy" (p. 141). However, he went on to claim that the Gaian system has maintained conditions comfortable for life nevertheless, with long-term homeostasis.

What if homeorrhetic behavior devolves into "progressive" Gaia, that is, temporary steady states on an evolutionary path? This version of Gaia, under the rubric of geophysiology, is just what I have argued for in recent papers and in this book (see chapter 8). In progressive Gaia, the biota mediates, but both the biota and biosphere coevolve. No real optimization for the biota occurs at any time. In my view, the fundamental challenge for the geophysiological Gaian research program is the search for self-regulation of other global effects of biotic/biospheric evolution besides the variation of atmospheric oxygen levels, for example, surface temperature, atmospheric composition, and self-regulatory mechanisms operating at smaller scales in the biosphere's subsystems. This question will be readdressed in chapter 9.

A Gaian Mechanism on Today's Earth?

One possible example of biological mediation today is the heating effect on the surface world ocean by virtue of phytoplankton albedo decrease (Sathyendranath et al. 1991). This decrease occurs as a result of pigment concentration. These researchers calculated a 4°C increase per month in late summer in the Arabian Sea from this effect. Biological processes thereby influence the transport of heat in the world ocean.

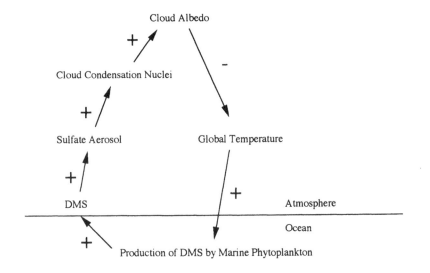

FIGURE 1-4.
Proposed feedback cycle between climate and marine DMS production. The plus and minus signs indicate if an increase in the value of the parameter is expected to lead to an increase (+) or decrease (−) in the value of the parameter at the arrow's end.

The dimethyl sulfide (DMS)-cloud feedback scenario leading to lower temperatures is another, a direct result of heuristic Gaia. DMS, produced by marine algae, oxidizes in the atmosphere to form sulphate aerosols that are cloud condensation nuclei (CCNs) over the ocean, raising cloud albedo and leading to surface cooling. The possibility of a negative feedback loop was originally proposed, where DMS production varies directly with temperature (figure 1-4). This Gaian scenario was proposed in two seminal papers (Charlson et al. 1987, with Lovelock as second author, and Shaw 1987), stimulating a whole research program and international meetings, with implications for anthropogenic sulphate aerosol cooling/global warming as well as biogeochemical issues related to algae and climate.

In an attempt to demonstrate how natural selection can potentially generate planetary self-regulation, Hamilton and Lenton (1998) proposed that DMS production may be linked to increased algal dispersal by promoting cloud formation. Does the proposed mechanism really work? Given the diversity of marine algae and complications in their response to temperature, it is still uncertain whether the net effect of the feedback is negative or positive. Recently, Lovelock and Kump (1994) (see further discussion in Kump and

Lovelock 1995) proposed that the overall feedback may be positive except during the coldest glacial episodes. The key feedback is the apparent positive link between cold water and marine productivity and DMS production. This postulated scenario is consistent with some evidence for higher sulphate aerosol concentrations during glaciations. A recent blow to negative feedback has emerged from a 15-year record of DMS in tropical Pacific waters that show little variation, despite big El Niño–related temperature and cloud variation over this period (Bates and Quinn 1997). If negative feedback is ultimately disproved in this case, it would be of little consequence given the enormous fruitfulness of the "error" in enlarging our knowledge.

Another outcome of the DMS-related research is the recognition that DMS-derived sulphate aerosols may dominate the non-anthropogenic CCN budget (Charlson 1991). Only eucaryotic algae apparently produce DMS, with the possible exception of some freshwater and perhaps even marine cyanobacteria (Hamilton and Lenton 1998). Assuming the absence of DMS production in the early Precambrian, without comparable alternative sources of CCNs, the Earth's albedo might have been significantly lower, approaching a cloud-free value of around 0.1. This effect alone would have resulted in higher surface temperatures. Model calculations of this scenario will be presented and discussed in chapter 8, since they are relevant to the issue of Precambrian surface temperatures.

A recent hypothesis has revived a Gaian DMS-related feedback, this time in a geophysiological model (Klinger and Erickson 1997). High coastal oceanic productivity is postulated to arise from the coupling of peatland and marine ecosystems. DMS production in marine coastal areas is supplying sulfur and organic acids to adjacent peatlands, enhancing their growth. In turn, organically complexed iron, a limiting nutrient in marine ecosystems, is enhancing phytoplankton productivity and thereby DMS production. This is a positive feedback loop. The authors find observational support in the association of peatland cover with adjacent marine chlorophyll concentration from a global survey.

In the next chapter we plunge into consideration of the carbon biogeochemical cycle, which begins the systematic consideration of my theory of biospheric evolution over geologic time.

Vernadsky and biogeochemistry. The biogeochemical cycle of carbon; today's cycle and its centrality in the greenhouse debate (a comparison of the anthropogenic and natural fluxes into and out of the atmosphere). The cycle on a geological time scale: the weathering and organic carbon burial sinks, the volcanic/metamorphic source. Another sink? (oceanic basalt reaction with seawater).

Vernadsky and Biogeochemistry

Vladimir Vernadsky's (1863–1945) enormous if somewhat premature contribution to science is still not fully appreciated in the West (Vernadsky 1998, originally published in Russian 1926; Ghilarov 1995; McMenamin and Mc-Menamin 1994; Bailes 1990; Westbroek 1991). In Russia, his homeland, Vernadsky was officially lionized in Soviet times, although he was far from being a staunch Marxist-Leninist. Ironically, it is now chic to disparage his stature. Vernadsky should be regarded as the father of biogeochemistry, having coined the word in 1926 in his book on the biosphere (Vernadsky 1945). For Vernadsky, the heart of biogeochemistry, the intersection of the biological, geological, and chemical realms, is the cycling of elements through the biosphere. The biosphere is seen as "a definite geological envelope markedly distinguished from all other envelopes of our planet. This is not only because it is inhabited by living matter, which reveals itself as a geological force of immense proportions, completely remaking the biosphere . . . but also because the biosphere is the only envelope of the planet into which energy permeates in a notable way, changing it even more than does living matter" (Vernadsky 1944). We even find the germ of homeostatic Gaia: "Living matter (as the biosphere itself) is organized in a way that is conducive to the function of the biosphere" (quoted in Ghilarov 1995 from Russian text).

A highly relevant thesis of Vernadsky to the central theme of this book is his statement that "the process of the weathering of rocks is a bio-inert process." He wrote, "It seems to me that this neglect explains the backwardness of the branch of chemical geology concerned with the zone of weathering, as constrasted with our present general level of knowledge. The process is approached as a physico-chemical one. A biogeochemical approach ought to contribute greatly to the solution of the problem" (Vernadsky 1944). Vernadsky defined bio-inert natural bodies as "regular structures consisting simultaneously of inert and living bodies," with a bio-inert process involving the interaction of these bodies.

Vernadsky apparently had a profound influence on the subsequent development of biogeochemistry in the West through a curious connection: his son George was a professor at Yale and a friend of G. E. Hutchinson, an influential force in the post–World War II development of ecology, biogeochemistry, and limnology. Hutchinson edited and introduced Vernadsky's first major publication in English, "Problems of Biogeochemistry II" (Vernadsky 1944) and also arranged publication of his paper in *American Scientist* in 1945 (Vernadsky 1945). In particular, there is strong resonance between Hutchinson's 1954 paper on the biochemistry of the terrestrial atmosphere, emphasizing biogeochemical interactions, and Vernadsky's concepts (Grinevald 1988). The biochemistry of the atmosphere indeed! Hutchinson's terminology anticipates the most provocative metaphors of Lovelock's Gaia.

Introduction to the Carbon Biogeochemical Cycle

The cycling of carbon through the biosphere, its biogeochemistry, is of critical concern today in light of global warming and its actual and potential multifold feedbacks to society. The enhanced greenhouse is produced by the reradiation of infrared to the Earth's surface by the anthropogenic greenhouse gases, carbon dioxide being the largest trace gas contributor (water vapor actually accounts for most of the greenhouse effect, but its level is dependent on the independent variation of atmospheric carbon dioxide). Central to all the debate and projections is knowing where the carbon dioxide emitted to the atmosphere ends up, and how this pattern might change as global surface temperature increases. Thus, knowledge of the multifold fluxes in and out of the systems and subsystems of the biosphere and their temporal and spatial variation is critical.

A summary of the global carbon cycle is shown in figure 2-1. First, let us take a look at the natural fluxes. The total photosynthetic flux is about 170 Pg C/year (the prefix P stands for Peta, 10^{15}), 50 for marine biota, and 120 for terrestrial. This flux is almost exactly balanced by a respiration and decay flux of carbon back into the atmosphere/ocean pool, mainly as carbon dioxide. Only a small flux of organic carbon and carbonate of about 0.2 Pg C/year is buried, constituting a sink with respect to the atmosphere/ocean pool. This latter flux balances the net source of carbon to the atmosphere, namely the volcanic source (about 0.1 Pg/year) and an equal flux of carbon from the oxidation of organic carbon present in exposed terrestrial rocks (not shown in figure 2-1).

Turning to anthropogenic fluxes to the atmosphere, these sum up the carbon dioxide from fossil fuel burning (about 5 Pg C/year) and deforestation from the decay of organic carbon (1–2 Pg C/year). Note that this sum (6–7 Pg C/year) is some 60 times the natural flux from volcanism and accounts for the well-known rise of carbon dioxide in the atmosphere over the past 100 years and the enhanced greenhouse effect. One critical flux to the long-term carbon cycle is subsumed in that of river-borne material, the flux of bicarbonate and calcium/magnesium ions derived from the weathering of CaMg silicates on land. The latter consist mainly of the following minerals: plagioclase (an NaCa feldspar, an aluminosilicate), biotite (sheet silicate containing magnesium), pyroxenes (single-chain silicates), olivine (single-tetrahedra silicate), and amphiboles (double-chained silicates).

Now consider the inventories of carbon in each reservoir. Carbon in the crust occurs mainly in the form of limestone and its metamorphic product marble and is some 500 times the mass of the total carbon in the atmosphere, biosphere, and ocean combined. [Note that other sources give larger reservoirs of carbonate and kerogen carbon (reduced organic carbon in sediments), 77.5 and 14.2 × 10^6 Pg, respectively (Holser et al. 1988)]. Oceanic carbon, mainly as bicarbonate ions, is some 54 times the mass in the atmosphere, whereas soil carbon is some four times the atmospheric carbon mass. Although the terrestrial biomass is more than 1000 times that of the oceanic

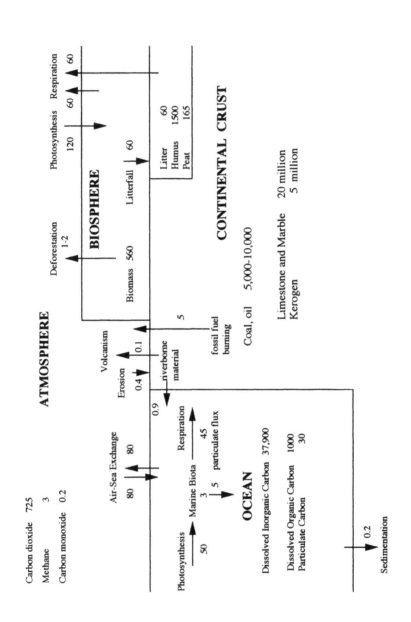

FIGURE 2–1.
A summary of the global carbon cycle. Reservoir contents are given in Pg (10^{15} g) and fluxes in Pg C/year (after Holmen 1992).

biomass, its rate of assimilation (photosynthesis) of carbon dioxide from the atmosphere is little more than twice the oceanic rate; most of the terrestrial biomass is in the form of dead wood in trees.

The difference between flux in and out and concentration in a reservoir needs to be clearly understood. Unfortunately, an influential recent book on environmental politics (Easterbrook 1995) confuses these two concepts, and thus is instructive:

> In absolute terms human-caused emissions of carbon dioxide have only increased the amount of this gas in the atmosphere by 0.006 percentage point. It's quite common to hear environmentalists express dismay over the 25 percent statistic [i.e., from 290 to 350 ppm], rare to find them bringing up the 0.006 side of the equation . . . Let's perform some simple manipulation of those numbers. Carbon dioxide constitutes roughly 1 percent of the full greenhouse effect, with the human-caused component of the carbon dioxide cycle at roughly 4 percent and the rate of artificial carbon dioxide increase around 1 percent annually. This works out to the human impact on the greenhouse effect being roughly 0.04 percent of the total annual effect. That is, 99.96 percent of global warming is caused by nature, 0.04 percent is caused by people. The present rate of increase in human-caused greenhouse impact, meanwhile, works out to about 0.004 percent per annum of the total effect (pp. 22–23) (Easterbrook 1995).

First, the carbon dioxide level during the ice ages was on the order of 180 ppm or 50% of the present level. Easterbrook's calculation of 0.006% is contrived. What Easterbrook fails to point out is that the water greenhouse is not the independent variable, but the carbon dioxide greenhouse is (i.e., the human impact is the full increase in the greenhouse, water, and carbon dioxide, as a result of anthropogenic carbon dioxide rise).

Finally, Easterbrook claims the following: "Compare the 20 percent of the atmosphere that is oxygen produced by land and sea vegetation against the 0.006 percent of the atmosphere that is carbon dioxide produced by human action. By this comparison, green plants put roughly 3,000 times as much gas into the atmosphere as power plants" (p. 24). Here Easterbrook confuses concentration with flux. Comparing fluxes, photosynthesis pumps

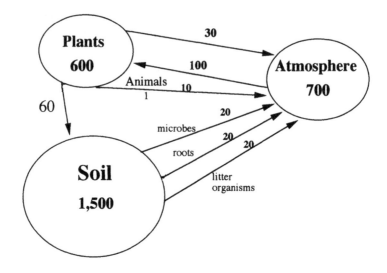

FIGURE 2-2.

Carbon fluxes and the soil. The amount of carbon in each reservoir is in Pg (10^{15} g), the flux in Pg/year. Steady state is assumed. The relatively small flux of carbon from animals to soil is not shown. (After Volk 1994.)

in 120 Pg C/year (and thereby releases an equivalent number of moles of oxygen, 10 Pmoles/year) or roughly 24 times the flux into the atmosphere from fossil fuel burning. However, the photosynthetic flux of oxygen is almost exactly balanced by the respiration and decay flux (the small burial of organic carbon is thus an oxygen source balancing natural sinks such as oxidation of ferrous iron in rocks). Recall that the flux of fossil fuel burning is about 50 times that of the natural flux of volcanic/metamorphic release of carbon dioxide to the atmosphere!

Zeroing in on one subsystem of the carbon biogeochemical cycle, the soil pool, we see a large potential positive feedback of global warming, the release of soil carbon into the atmosphere (figures 2-1 and 2-2). Note that the ratio of soil organic carbon to atmosphere carbon is about 2:1. The global estimate of soil organic matter divided by the carbon deposited as litter gives a mean residence time of about 25 years. However, the residence time is apparently significantly reduced as temperature increases (Trumbore et al. 1996), with big releases of soil carbon to the atmosphere expected from global warming.

What Controls the Long-term Climate?

Although on short time scales of less than 10^4 years the cycling between the atmosphere/ocean and surface pools such as organic carbon can have significant impact on atmospheric carbon dioxide levels (witness the glacial/interglacial cycles of the last 2 million years, anthropogenic impacts, etc.), the long-term cycle ($>10^5$ years) is controlled by the silicate-carbonate geochemical cycle. This cycle entails transfers of carbon to and from the crust and mantle. In the modern era, this cycle was first described by Urey (1952):

$$CO_2 + CaSiO_3 = CaCO_3 + SiO_2$$

The reaction to the right side of the equation corresponds to chemical weathering of calcium silicates on land ($CaSiO_3$ is a simplified proxy for the diversity of rock-forming CaMg silicates such as plagioclase and pyroxene, which have more complicated formulas, e.g., calcium plagioclase, $CaAl_2Si_2O_8$), whereas the reaction to the left corresponds to metamorphism (decarbonation) and degassing returning carbon dioxide to the atmosphere. The main aspects of chemical weathering, including even the realization that plants are accelerators, and the long-term control mechanism on carbon in the atmosphere were first published over 140 years ago by the French mining engineer Jacques Ebelmen (Berner and Maasch 1996). Ebelmen anticipated with great lucidity the silicate-carbonate geochemical cycle later described with less completeness by Hogbom 40 years later (Berner 1995b). Hogbom left out the decarbonation source of carbon dioxide, which amazingly Ebelmen included. Apparently these ideas were forgotten and as so often is the case in the history of science, the long-term carbon cycle was rediscovered by Urey in 1952.

This cycle is really biogeochemical. Although decarbonation and outgassing is surely abiotic (taking place at volcanoes associated with subduction zones and oceanic ridges), chemical weathering involves biotic mediation. This aspect, as well as the mechanism for negative feedback control of the carbon dioxide level in the atmosphere/ocean system, will be discussed more fully in chapter 3. For now, it is sufficient to emphasize that chemical weathering requires a flow of water and carbon dioxide through a layer of soil, with a high reactive surface area of CaMg silicates if consumption of atmospheric carbon dioxide is to occur at a rate similar to that on today's Earth. Thus, most chemical weathering occurs on vegetated continental surface in temperate and tropical climates because of moderate to high rainfall and temperatures. Naturally, higher temperatures, with other conditions constant, mean higher rates of reaction. In general, rocks that formed at high temperatures from cooling magma, such as basalt, weather fastest with respect to atmospheric carbon dioxide consumption. Putting carbon dioxide and water into the weathering equation for a common rock-forming mineral in basalt, calcium-rich plagioclase (simplifying by leaving out the sodium component):

$$CaAl_2Si_2O_8 + 3H_2O + 2CO_2 \rightarrow Ca^{2+} + 2HCO_3^-$$
$$+ Al_2Si_2O_5(OH)_4$$

$(Al_2Si_2O_5(OH)_4$ is kaolinite, a clay mineral)

The products include dissolved calcium ion and bicarbonate ion, which ultimately wind up in the ocean from river input, and kaolinite, left in the soil.

Steady-state levels of carbon dioxide in the atmosphere are achieved on time scales on the order of 10^5 to 10^6 years (Sundquist 1991). As a first approximation, the time needed to reach steady-state levels is the residence time of carbon in the atmosphere/ocean pool with respect to the volcanic source, or $(40,000/0.1) = 4 \times 10^5$ years. Note that carbon rapidly equilibrates within the atmosphere/ocean pool ($\leq 10^3$ years), with about 54 times as much carbon in the present ocean as in the atmosphere.

That the residence time is a measure of the feedback strength, or time to reestablish steady state, is illustrated by a leaky bucket model (thanks to Tyler Volk for this analogy). Let a faucet supply water to a bucket with a hole allowing outflow. Assume a steady-state level of water is reached in the bucket, analogous to the carbon dioxide level in the atmosphere/ocean pool. Because the residence time of water in the bucket is inversely proportional to the inflow rate, increasing the residence time for the same water level means the inflow rate must be lower, thereby taking longer to reach a new steady state.

Perturbations from the carbon dioxide steady state in the atmosphere/ocean pool can occur as a result of Earth orbital variations, fluctuations in organic carbon burial, pulses in volcanic outgassing, and so forth. One way to describe the perturbation from steady state is the e-fold response time (Rodhe 1992), that is, the time to restore the mass of something in a reservoir (in this case the carbon dioxide level in the atmosphere) to $1/e = 37\%$ of the initial imbalance. Sundquist's (1991) modeling give e-fold response times of 300 to 400×10^3 years for the near present-day carbonate–silicate cycle.

Sources and Sinks: The Canonical Equality

A steady-state carbon dioxide level in the atmosphere/ocean system is achieved by the equality of the sink fluxes (removal processes) and the sources (supply processes) to this pool. At a given time the partitioning of carbon between atmosphere and ocean is determined by global temperature (or the corresponding atmospheric pCO_2) and either the pH or degree of carbonate saturation of the ocean (i.e., the carbonate and bicarbonate level). Now most of the carbon in the atmosphere/ocean pool is in the ocean (see figure 2-1) with a ratio of atmospheric to oceanic carbon of 1:54 (in the Archean/early Proterozoic, assuming atmospheric pCO_2 levels of about 1 bar, the ratio was likely closer to 1:1 because the speciation of carbon in the ocean shifts from bicarbonate to dissolved carbon dioxide with rising pCO_2 and the solubility of carbon dioxide in water is relatively small). As we will see in a later discussion, the atmospheric pCO_2 level can be computed as a first approximation from modeling the long-term steady state as a function of variation in the volcanic/metamorphic source and silicate weathering sink, the latter determined by biotic influence and land area.

For the atmosphere/ocean pool to remain at steady-state on a time scale of 10^5 to 10^6 years or more, the sums of the input fluxes must equal the output fluxes (figure 2-3). F_1 corresponds to the flux of carbonate deposited in the ocean, derived from the reaction of atmospheric carbon dioxide with CaMg silicates and carbonates in weathering reactions on land. F_2 is the flux of carbon from the dissolution of land carbonates alone. Thus $(F_1 - F_2)$ corresponds to the CaMg weathering sink alone because a flux equal to F_2 is deposited as carbonate in the ocean. F_3 is the flux of organic carbon into the sedimentary reservoir from the net deposition in the ocean from both terrestrial and oceanic sources. F_4 is the carbon flux back into the pool derived from the weathering of organic carbon in exposed sedimentary rock on land (e.g., oxidation of coal). V is the volcanic/metamorphic carbon source flux. Then $V = (F_1 - F_2) + (F_3 - F_4)$. This is the canonical equality for a steady-state carbon dioxide level in the atmosphere/ocean pool.

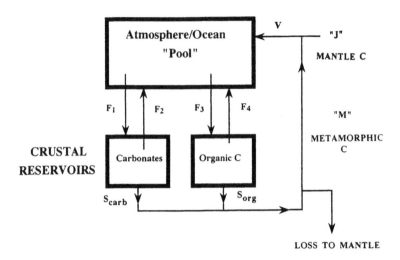

FIGURE 2-3.
Modeling the steady-state level of carbon in the atmosphere/ocean pool. F_1 is the flux corresponding to the deposition of carbonate, F_3 the flux of deposition of organic carbon in sediments, F_2 and F_4 the sedimentary recycling fluxes of carbonates and organic carbon, respectively, and S_{carb} and S_{org} the subduction fluxes to the mantle of carbonate and organic carbon, respectively. V is the total flux of carbon degassed to the atmosphere/ocean, subdivided into a recycled metamorphic flux M and a juvenile flux J.

The evolution of carbon sinks with respect to the atmosphere/ocean reservoir over geologic time remains uncertain despite inferences made from the sedimentary carbon isotopic record (Schidlowski 1988; Schidlowski and Aharon 1992; Des Marais et al. 1992; Holser et al. 1988). Making implicit assumptions explicit in such discussions should clarify the interpretation of published research and suggest new approaches to better understand the Earth's long-term regulator of surface temperature, the carbonate–silicate geochemical cycle, particularly the role of biota (Berner 1992; Schwartzman and Volk 1989; Schwartzman and Caldeira 1995). This subject is discussed in the appendix.

Other Sinks?

Low temperature alteration of seafloor basalt by seawater, leaching calcium, and then precipitating calcium carbonate has been proposed as an additional carbon sink with respect to the atmosphere/ocean pool (Staudigel et al. 1989; Francois and Walker 1992; Brady and Gislason 1997). Note that this sink would be exactly balanced by a corresponding volcanic/metamorphic source if complete decarbonation and release of carbon dioxide to the atmosphere occurs in subduction zones (Berner 1990b; Staudigel et al. 1990; see discussion of this issue in chapter 7). In a comprehensive review of this proposed sink, Caldeira (1995) concluded that silicate weathering on land and not low temperature seafloor basalt alteration has been the primary control on long-term atmospheric carbon dioxide levels over geologic time. This conclusion follows from the much greater sensitivity of silicate weathering on carbon dioxide levels than the proposed seafloor weathering sink. Brady and Gislason (1997) also agree that seafloor weathering is less responsive to carbon dioxide levels than silicate weathering on land. However, an intriguing possibility is that higher seawater temperatures might accelerate microbially mediated dissolution of seafloor basalts, thereby constituting a strong negative feedback (for the evidence of microbial dissolution, see Staudigel et al. 1995 and Thorseth et al. 1995a, 1995b).

A related issue is the flux levels of alkali and alkaline earths to and from hydrothermal brines that have interacted with seafloor basalt near mid-ocean ridges (Elderfield and Schultz 1996). Some researchers think calcium release to seawater is stoichiometrically balanced by magnesium uptake to form spilitic (sodium-enriched) greenstone and amphibolite (Berner and Berner 1987). If this is the case, then the weathering sink for carbon involves only weathering of silicates on land. If the cation balance for magnesium uptake involves both sodium and calcium ions (Spencer and Hardie 1990; Hardie 1996) then some bicarbonate derived from land weathering of sodium silicates (e.g., plagioclase) could be precipitated as calcium carbonate in the ocean, thus constituting an additional carbon sink (Spivak and Staudigel 1994). A major uncertainty is the contribution of fluxes through ridge flanks, where seawater circulates through cooling oceanic crust as it spreads from the ridge (Elderfield and Schultz 1996).

Complicating this interpretation is the possibility of significant reverse weathering occurring in the ocean (Mackenzie and Kump 1995; Michalopoulos and Aller 1995). Reverse weathering involves the formation of clay minerals in oceanic sediments from the reaction of silica, degraded aluminum clays, iron oxide, organic carbon, soluble cations, and bicarbonate in new shore oceanic sediments forming new clay with a net release of carbon dioxide (Mackenzie and Kump 1995). The significance of these reactions to the carbon cycle is still unclear; results to date support uptake of some 10% of the K^+ input to the ocean. In any case, if the land weathering of NaK silicates turns out to be a contribution to the carbon sink with respect to the atmosphere/ocean pool, it has similar kinetics to CaMg silicate dissolution and is likely to be of secondary significance to the weathering of CaMg silicates, and is not of major importance in modeling the long-term carbon cycle (Berner, personal communication).

Appendix: Rethinking the Sedimentary Carbon Isotopic Record

INTRODUCTION

The interpretation of the history of the carbon biogeochemical cycle is more uncertain than commonly assumed. The main source of this uncertainty is the possible variation in the isotopic composition of degassed carbon and relative fluxes of deposited and weathered carbon over geologic time. Furthermore, preserved sedimentary inventories of kerogen and carbonate probably give a misleading picture of the history of their relative deposition when naively interpreted. Knowledge of the relative importance of the silicate weathering versus the net organic carbon sink with respect to the atmosphere/ocean system is needed to better understand the evolution of Earth's climate.

The main ambiguity in the published literature alluded to above has centered on the interpretation of the simple isotopic mass balance:

$$\delta_{in} = \delta_{org} f_{org} + \delta_{carb} f_{carb},$$

where the δ's represent the $\delta^{13}C$ values for the input, organic carbon burial, and carbonate carbon burial fluxes, respectively, at a given time. $\delta^{13}C$ is defined as $= [(^{13}C/^{12}C)_{sample}/(^{13}C/^{12}C)_{standard} - 1] \times 1000$ where $\delta^{13}C$ is in parts per thousand (‰). The standard has a $^{12}C/^{13}C = 88.99$. δ_{in} can be variously interpreted as representing the carbon flux to the atmosphere/ocean pool from (i) mantle carbon alone (J in figure 2-3); (ii) mantle carbon, plus carbon from magmatism and metamorphism of sedimentary rock (M in figure 2-3); or (iii) J plus M plus carbon from sedimentary recycling of carbonate and organic carbon (F_2 and F_4, respectively, in figure 2-3) (Des Marais et al. 1992;

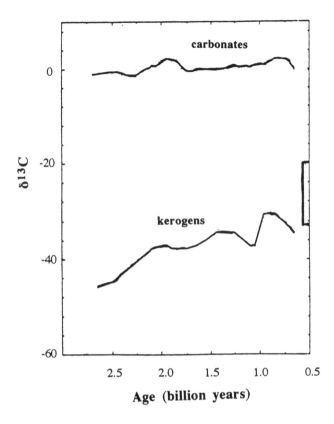

FIGURE 2-4.
The lines depict running averages of δ¹³C values of carbonates and purified kerogens (corrected for postdepositional alteration) as a function of age. The rectangle along the right margin shows the range of Phanerozoic $\delta^{13}C$ kerogen values. (After Des Marais et al. 1992.)

see figure 2-4). In each case, f_{org} and f_{carb} represent the burial fraction of the input carbon flux that goes to organic carbon and carbonate sedimentary reservoirs, respectively, at a given time.

For interpretation (iii) the carbon flux associated with f_{carb} corresponds to the total deposition of carbonate (F_1), that is, recycled older carbonate in addition to carbonate derived from the weathering of silicate rock. f_{org} likewise corresponds to the total deposition flux of organic carbon (F_3). Thus, for interpretation (iii):

$$V + F_2 + F_4 = F_1 + F_3$$

For interpretations (i) and (ii), the carbon flux associated with $f$carb includes only silicate rock weathering (and subsequent carbonate burial), $(F_1 - F_2)$, that associated with $f_{org,}$ $(F_3 - F_4)$, since in these cases $V = (F_1 - F_2) + (F_3 - F_4)$, f_{carb}/f_{org} is the ratio of the silicate weathering sink to the net organic carbon burial sink. Unless δ_{in} for cases (i) and (ii) are assumed equal [with no metamorphic flux assumed for case (i)], the computed f_{carb} and f_{org} values are different for cases (i) and (ii). Furthermore, for the canonical equality case,

$$\delta_{org} = f_{F_3}\delta_{F_3} - f_{F_4}\delta_{F_4}$$

Similarly,

$$\delta_{carb} = f_{F_1}\delta_{F_1} - f_{F_2}\delta_{F_2}$$

In the isotopic mass balance, δ_{in} has been assumed to correspond to $-5‰$, the presumed juvenile $\delta^{13}C$ (Schidlowski and Aharon 1992). A value of $-5‰$ is near the diamond mode of $-5.5‰$ (Deines 1992), and close to the inferred weighted average of the extant crustal reservoirs of oxidized and reduced carbon (Holser et al. 1988). However, there is no strong reason to assume that this value is necessarily correct for any of the above interpretations, including case (i). The $\delta^{13}C$ of mantle carbon is apparently uncertain, with cogent arguments having been made for a mantle with inherited initial heterogeneity of carbon isotopic composition ranging from 0 to $-30‰$ with a weighted mean of $-7‰$ (Deines 1992). Others have interpreted this variability as evidence for subducted biogenic carbon (Kirkley et al. 1991). Roughly one sixth of the present volcanic/metamorphic carbon dioxide flux is released by Mt. Etna (Brantley and Koepenick 1995) with a $\delta^{13}C = -3.7$ (Allard et al. 1991; Gerlach 1991), essentially identical to the initial magmatic value reported from the mid-ocean ridge basalts (MORB) (Pineau and Javoy 1994). Sano and Williams (1996) assumed a mantle value of -6.5 corresponding to MORB in their calculations. In any case the jury is still apparently out on this issue.

Thus, for cases (ii) and (iii), there is no a priori reason to assume $\delta_{in} = -5‰$ or that δ_{in} has remained constant over geologic time. Various histories may have affected δ_{in}, namely that of carbonate sedimentation (e.g., onset of pelagic carbonate sedimentation in the Cenozoic; see Volk 1989a and Caldeira 1991), the change in depth of the source of mantle-derived magmas and volatiles, and change in styles in plate tectonics. For example, from the observed carbon and helium isotopic compositions and $CO_2/^3He$ ratios of volcanic gases, Sano and Williams (1996) estimated that 60% of the present degassed carbon comes from subducted sediment carbon (mainly carbonate).

SEDIMENTARY INVENTORIES AND CARBON SINKS

Another approach to understanding the history of the carbon biogeochemical cycle has been to inventory the masses and weighted averages of $\delta^{13}C$ of the preserved sedimentary reservoirs. Holser et al.'s (1988) compilation (taken from Galimov et al. 1975) gives $\delta^{13}C = -4.5\permil$ for the total crustal inventory, and a carbonate carbon to organic carbon molar ratio of 5.5. Because what is preserved represents the surviving net sinks from the atmosphere/ocean pool, the latter value corresponds to the ratio of the cumulative silicate weathering to organic carbon sink if no differential destruction (via metamorphism, subduction, etc.; shown as S_{carb} and S_{org} in figure 2–3) has occurred since the onset of the creation of the sedimentary reservoirs as demonstrated below:

Let m_{carb} and m_{org} be the crustal masses of carbonate and organic carbon respectively at any time t (age).

$$\text{Let } F_a = (F_1 - F_2), \text{ and } F_b = (F_3 - F_4)$$

Then,

$$dm_{carb}/dt = Fa(t) - S_{carb}(t) \text{ and } dm_{org}/dt = F_b(t) - S_{org}(t)$$

Then,

$$m_{carb}/m_{org} = \int F_a(t)\, dt / \int F_b(t)\, dt$$
$$\text{if } R = \int F_a(t)\, dt / \int F_b(t)\, dt = \int S_{carb}(t)\, dt / \int S_{org}(t)\, dt$$

where R is the ratio of the cumulative silicate weathering sink to the cumulative organic carbon sink. How realistic is the requirement for no differential destruction? Because the preserved inventories are heavily dominated by younger sediments (Veizer 1988), any temporal change in the relative proportions of carbonate and organic carbon deposition going back into the Precambrian would be suppressed in the extant record, thereby creating an effective differential destruction violating the required condition for $R = m_{carb}/m_{org}$.

However, even if there has been no differential destruction of the two crustal reservoirs, the weighted mean $\delta^{13}C$ of the composite of the surviving reservoirs is not necessarily equal to the $\delta^{13}C$ corresponding to the cumulative juvenile and metamorphic flux because the preserved inventories are heavily biased by younger sediments and the $\delta^{13}C$ of organic carbon has become progressively less negative to the present (using the data set corrected for burial effects given in Des Marais 1992; this effect was not taken into account in Holser et al. 1988).

For interpretation (ii), $f_{org} = (F_1 - F_2)/V$, corresponding to the net organic carbon sink with respect to the atmosphere/ocean system. Since both V and F_1 were plausibly greater in the Archean/early Proterozoic, f_{org} may have been roughly constant over geologic time. A higher F_1 in the early Precambrian is consistent with suggestions that higher atmospheric carbon dioxide levels then resulted in greater oceanic productivities than now (Rothschild and Mancinelli 1990) and possibly greater organic carbon burial rates. Perhaps in the early Archean, high marine productivities were associated with hydrothermal vent sites (de Ronde and Ebbesen 1996). This conjecture may be supported by the apparently isotopically heavier kerogen record up to about 3 Ga.

If δ_{in} is interpreted as in case (ii) and has been nearly constant since the early Precambrian, and assuming $\delta^{13}C$ for F_1, F_2 and F_3, F_4 fluxes correspond to the measured values of preserved carbonate and organic carbon, respectively, then the inversion of the sedimentary carbon isotopic record (Des Marais et al. 1992) gives a higher inferred ratio of the silicate weathering to the organic carbon sink in the Precambrian (especially the Archean) than now:

$$f_{org} = (\delta_{in} - \delta_{carb})/(\delta_{org} - \delta_{carb})$$

At 3 Ga, assuming $\delta_{in} = -5$, $\delta_{org} = -40$, $\delta_{carb} = 0$, gives $f_{org} = 0.125$ compared with f_{org} now $= 0.2$. If this interpretation is valid, then this trend is consistent with a lower biotic productivity on land in the Precambrian, and lower organic carbon burial derived from terrestrial sources relative to the case for continents colonized by higher plants in the Phanerozoic (McMenamin and McMenamin 1994). A higher ratio of the silicate weathering to the organic carbon sink in the Precambrian also implies that the molar ratio of 5.5 inferred from the extant record is a minimum. Presently, the ratio of silicate to net organic carbon sinks, that is, $(F_1 - F_2)$ to $(F_3 - F_4)$, is estimated to be about 5 to 6 (from data in Berner 1991). The present ratio of the organic carbon burial flux (F_3) to weathering of kerogen flux (F_4) is roughly 1.3 (data from Berner 1991). If the weathering flux is related to the level of atmospheric pO_2 (Berner 1989) then this flux might have been lower during the Precambrian, even with higher physical erosion rates. The completeness of remineralization of kerogen in weathering is uncertain (Hedges 1992). Furthermore, it is possible at times that $F_4 > F_3$, that is, the flux of carbon dioxide from oxidation of kerogen exceeds the burial flux of organic carbon, the net effect being a source of carbon dioxide to the atmosphere/ocean pool (Beck et al. 1995).

Following interpretation (iii), δ_{in} may have been more variable than for (i) or (ii) because δ_{in} reflects the combined fluxes $V + F_2 + F_4$. Is the kerogen record consistent with a constant f_{org} as defined by interpretation (iii), with f_{org} here corresponding to the actually measured values in preserved kerogen of a given age (the data set filtered for burial effects)? We give a tentative positive answer because relatively small variations in δ_{in} and δ_{carb} can compensate for the large variation in δ_{org}, illustrated as follows:

Assuming at 3 Ga $\delta_{org} = -40$, $\delta_{carb} = 0$, for $f_{org} = 0.2$ requires $\delta_{in} = -8$. A more negative δ_{in} in the Archean is consistent with a smaller contribution of subducted carbonate and/or temporal variations in mantle $\delta^{13}C$ arising from reinjection of recycled carbonate. The observed trend in $\delta_{org} - \delta_{carb}$ has been plausibly explained as a result of greater atmospheric carbon dioxide levels in the Precambrian (Strauss et al. 1992a) as well as a greater role of methanotrophy in the late Archean (Hayes 1994).

In conclusion, the actual state of knowledge of the partitioning of carbon burial and ratio of silicate weathering to organic carbon sinks with respect to the atmosphere/ocean over geologic time is more uncertain than previously assumed. Perhaps greater insight will arise from a more detailed inventory of both the masses and $\delta^{13}C$ of sedimentary carbon preserved in the geologic record as well as data bearing on the possibility of differential destruction of carbonate versus organic carbon reservoirs during subduction and/or metamorphism.

The standard model of solar luminosity variation over geologic time and the
faint young sun paradox. Recent challenges to the standard model. The silicate-
carbonate climatic stabilizer (**WHAK** model). Modeling the silicate–carbonate climatic stabilizer from the ground up. From **BLAG** to **GEOCARB**: modeling the Phanerozoic. Phanerozoic atmospheric carbon dioxide levels.

The Standard Model of Solar Luminosity Variation Over Geologic Time and the Faint Young Sun Paradox

Sagan and Mullen (1972) proposed that the sun's luminosity (energy flux)
was probably low enough to freeze the oceans some 2 billion years ago if the
atmospheric composition were the same as now (the faint young sun paradox), yet for the past 3.5 billion years the Earth has had liquid oceans, a fact
established from the fossil and marine sedimentary record (figure 3-1). Sagan
and Mullen's solution to this paradox was the presence of a reducing gas such
as ammonia in the early atmosphere, as a greenhouse gas in sufficient levels to
preempt freezing. This solution was later rejected because of the computed
instability of the proposed gases but recently has been reproposed. The most
likely solution of the faint young sun paradox is that atmospheric compositions, particularly of greenhouse gases, have not been constant over geologic
time. Hence, some mechanism for regulation of surface temperature through
geologic time is apparently needed. The most likely candidate for this role is
the greenhouse gas carbon dioxide, although greenhouse contributions from
traces of methane and/or ammonia in the atmosphere should not be ruled
out. This issue will be addressed more thoroughly in chapter 7.

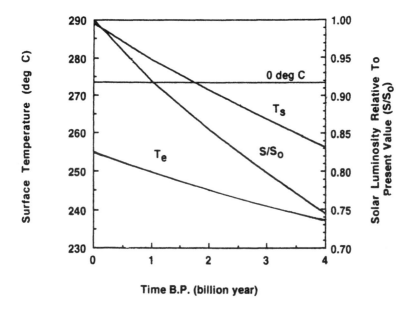

FIGURE 3-1.
The faint young sun paradox. Curve *Te* is the effective radiating temperature of the Earth (constant planetary albedo of 0.3). Curve *Ts* is the surface temperature of Earth for a constant atmospheric composition (same as now) [from Kasting and Ackerman's (1986) climatic model].

Recent Challenges to the Standard Model

A standard model of luminosity (L) variation of the sun (Sackmann et al. 1990) has been assumed in my simulations of the long-term climate of Earth. It is generally accepted that the zero age main sequence (ZAMS) solar luminosity, at the initial stage of core hydrogen burning, was about 70% of the present value and the sun's surface temperature was about 3% lower (Bahcall et al. 1982). Although a steadily increasing energy flux over time might seem counterintuitive, it results from the increase in the rate of fusion of H to He in the sun's core as it heats up as it contracts. However, some reports in the literature account for a warm early Earth without the intervention of a greenhouse effect. One of the most prominent is the discussion by Graedel et al. (1991) of a more luminous early sun (a similar argument is made by Doyle et al. 1993). The argument assumes that the mass of the sun was higher during roughly the first billion years of its tenure on the ZAMS. Subsequent mass loss is presumed to have reduced the mass to its present value. Because of the strong dependence of solar luminosity and radius on the star's mass, if the sun were about 10% more massive, it would have been about 50% brighter and about 10% hotter (note that surface temperature and energy flux are not directly proportional). This, they argue, is sufficient to completely account for the initially warm climate of the planet. The idea follows a more general suggestion by Willson et al. (1987) that this sort of mass loss is typical of stars when they initiate hydrogen core burning and slightly before.

However, the solar neutrino flux places severe constraints on any possible mass reduction after the onset on core nuclear processing. If the sun were indeed initially more massive, it would have a more massive core and an even larger deficit of neutrinos expected from fusion reactions than is presumed to be the case (Shore, personal communication). In addition, current models of the solar interior agree with the observed spectrum of surface and envelope oscillations (helioseismology) (Bahcall and Pinsonneault 1992). Thus, despite the claims made by Doyle et al. (1993), there is no empirical support for extended early main sequence mass loss. This scenario of early solar mass loss also has been offered as an explanation of warm temperatures on early Mars, implied by evidence for early liquid water on its surface (Whitmire et al. 1995). It is highly unlikely that the mass of the sun has changed appreciably during its lifetime (Shore, personal communication), and therefore the standard model of L variation has been used for modeling of the long-term carbon cycle.

The Silicate-Carbonate Climatic Stabilizer

In 1981, James Walker, P. B. Hays, and James Kasting proposed that temperature regulation in the silicate–carbonate geochemical cycle constituted a climatic stabilizer. This became known as the WHAK model after the names of its authors. This mechanism was derived from the Urey equilibrium discussed in chapter 2. To recapitulate, only the reaction of calcium and magnesium silicates with carbonic acid results in a carbon sink via the formation of bicarbonate, its transfer to the ocean, and the precipitation of $CaCO_3$; the weathering of limestone produces no net change in carbon dioxide in the atmosphere/ocean system, nor as a first approximation does the weathering of NaK silicates because no calcium or magnesium is supplied to the ocean (see discussion of other sinks in chapter 2).

On land:

$$CaSiO_3 + H_2O + 2CO_2 \rightarrow 2HCO_3^- + Ca^{2+} + SiO_2$$

In ocean:

$$2HCO_3^- + Ca^{2+} \rightarrow CaCO_3 \downarrow + CO_2 + H_2O$$
(reverse reaction is weathering of limestone on land)

Note that for each mole of $CaSiO_3$ reacting, there is a net consumption of one mole of CO_2, which is buried on the ocean floor as limestone. The deposition of limestone is of course largely biotically mediated, a process vividly described by Westbroek (1991). However, deposition of calcium carbonate would surely occur under abiotic conditions as well.

The heart of the stabilizer is the dependence of the weathering rate on surface temperature, itself controlled by carbon dioxide in the atmosphere. This rate increases with increasing temperature because of two main effects: the speed-up of chemical reactions and the increase in rainfall and therefore river runoff with increasing temperature. These effects are empirically supported from global studies of weathering rates and solute levels in rivers, as well as theoretically by models. Negative feedback resulting in temperature stabilization is obtained because the carbon sink increases as temperature (carbon dioxide) increases. But does life mediate this mechanism? If it does, what if any difference does it make? For example, is the temperature different because of biotic influences? We will return to this question in chapter 6.

Modeling the Silicate–Carbonate Climatic Stabilizer from the Ground Up

Modeling the silicate–carbonate climatic stabilizer is now considered first by assuming very simple, even unrealistic conditions, then adding increasing complexity (i.e., geologic and climatic realism) in successive scenarios of the evolution of climate, aimed at simulating the history of global mean surface temperatures of the past 4 billion years.

We begin with model A: solar luminosity varies according to the standard model, hence the energy flux to the Earth surface steadily increases to the

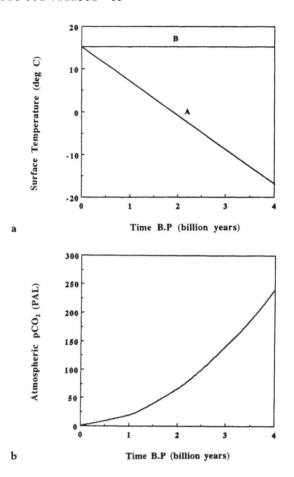

FIGURE 3-2.

a: Surface temperature as a function of time for two conditions: A, atmospheric pCO_2 level constant over time; B, volcanic outgassing of CO_2 (V) and land area constant over time. **b:** Variation of atmospheric pCO_2 level over time for model B.

present over geologic time. Model A assumes the carbon dioxide level in the atmosphere (pCO_2) is constant, the same as the present level of 3×10^{-4} bar, and no temperature dependence of weathering, hence no functioning of the carbonate-silicate cycle. The surface temperature (Ts) simply tracks the luminosity change, the temperature being computed using a standard greenhouse function, with $Ts = f(L, pCO_2)$. Because the energy flux was lower in the past, so was Ts for the same pCO_2 (figure 3-2a). Note that this temperature scenario is identical to that invoked for the faint young sun paradox (see figure 3-1).

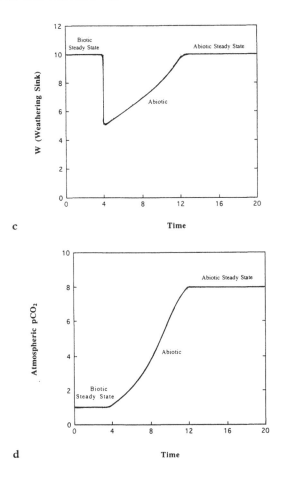

FIGURE 3-2 CONTINUED.
c: Weathering sink (flux) as a function of time, illustrating the effect of suddenly eliminating the biotic enhancement of weathering, that is, going from a biotic to an abiotic
land surface (units arbitrary). d: Same scenario as in c, but tracking the change in atmospheric pCO$_2$ level (units arbitrary).

Now let's allow the weathering rate to vary with temperature, bringing
into operation the silicate–carbonate climatic stabilizer at a minimal level.
Returning to the canonical equality (see figure 2-3), recall that V is the vol-
canic/metamorphic outgassing source of carbon; F_1 corresponds to the total
flux of carbonate deposited in the ocean, derived from both the reaction of
atmospheric carbon dioxide with CaMg silicates and carbonates in weathering on land; and F_2 is the flux of carbon into the atmosphere/ocean pool

from the dissolution of land carbonates alone. Thus, $(F_1 - F_2)$ corresponds to the CaMg weathering sink alone because a flux equal to F_2 is deposited as carbonate in the ocean. Ignoring F_4, the flux of weathering of organic carbon back into the atmosphere/ocean pool, as well as F_3, the burial of organic carbon, then,

$$V = F_1 - F_2$$

The volcanic/metamorphic outgassing source of carbon equals the silicate weathering sink, generating a steady-state level of carbon dioxide in the atmosphere. In this model (B in figure 3-2a), V and the area of continent available for silicate/carbonate weathering are assumed constant over time. The value of pCO_2 at present is assumed to be the same as now. Because V is assumed constant, the silicate weathering rate also must be constant over time. If this rate depends on temperature, then Ts must be constant. But because L varies, so must pCO_2. The computed variation of pCO_2 is shown in figure 3-2b.

If V is allowed to vary, then the silicate weathering sink must also vary over geologic time. For higher V in the past, plausible because of higher levels of uranium and ^{40}K in the solid Earth, producing more heat from radioactive decay, then the silicate weathering sink would also be higher at steady state. For this case, assuming temperature is the only parameter affecting weathering rate, higher surface temperatures in the past would be required and therefore even higher pCO_2 than for model B. Assuming some continental growth for the past 4 billion years and assuming that the silicate weathering sink is proportional to land area, then the effect of smaller continental land areas in the past would be to require still higher Ts and pCO_2 to generate the same weathering sink.

Now let's introduce some biology into this simple model. Suppose soil biota results in a higher silicate weathering rate than for an biotic continental land surface at the same atmospheric pCO_2 and surface temperature. Then the presence of soil biota allows the steady-state atmospheric pCO_2 and surface temperature to be lower than the abiotic case for the same required weathering flux (W), equal to V, which is assumed to be independent of surface conditions. Suppose the continental land surface is suddenly sterilized (figure 3-2c), eliminating the biotic effect on weathering. Then W should suddenly drop, destroying the steady-state condition. Because W is now less than V, then atmospheric pCO_2 and surface temperature will rise (figure

3-2d) until new steady-state atmospheric pCO_2 and surface temperature are achieved. Then W equals V, but at a higher atmospheric pCO_2 and surface temperature, now at abiotic conditions.

This modeling will be picked up again after a necessary detour, first considering the research program to model the past 600 million years, the Phanerozoic, then a much closer look at the weathering process and its biotic involvement discussed in chapters 4 through 6, which is central to my theory of biospheric evolution.

From BLAG to GEOCARB: Modeling the Phanerozoic

Soon after WHAK, the first attempt was made to model the long-term carbon cycle at least for the past 100 million years (Berner et al. 1983). This model became known as BLAG after the names of the authors of the original paper (i.e., Berner, Lasaga, and Garrels). In the first BLAG model, steady-state atmospheric pCO_2 levels (and surface temperature) were computed from the balance between continental weathering of carbonates and CaMg silicate sinks and the source, metamorphic/magmatic degassing derived from the decarbonation of subducted CaMg carbonates now and in the past assuming the following:

1. Continental land area as a function of time, because land area will affect the weathering sink; as a first approximation, the silicate weathering sink is directly proportional to land area.
2. Seafloor spreading rate as a function of time, because the volcanic outgassing rate is expected to vary with spreading rate.
3. Weathering rate as a function of temperature and runoff (itself a function of temperature).
4. Calcium and bicarbonate concentrations and calcium carbonate precipitation as a function of atmospheric pCO_2.

Eight differential nonlinear mass balance equations were solved simultaneously from assumed initial conditions at 100 million years before present (BP) (e.g., equations of the form: $dx/dt = -kx$, where x is the mass and k is the rate constant). There was no explicit inclusion of biological effects, particularly on weathering kinetics.

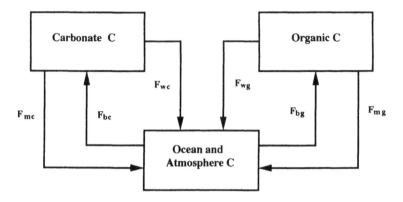

FIGURE 3-3.
The geochemical carbon cycle as interpreted by the GEOCARB model. F_{wc} is weathering flux of carbonates, F_{mc} the degassing flux from thermal carbonate decomposition, F_{wg} the weathering flux of organic matter, F_{mg} the degassing flux from thermal organic matter decomposition, F_{bc} the burial flux of carbonates in sediments, and F_{bg} the burial flux of organic matter in sediments. A steady state, required over a multimillion year time scale gives $F_{wc} + F_{mc} + F_{wg} + F_{mg} = F_{bc} + F_{bg}$ (after Berner 1995a).

For the next decade, BLAG evolved with greater complexity, extending the time period back over the whole of Phanerozoic time (Berner and Barron 1984; Berner 1990a, 1991, 1994; for a popularization, see Berner and Lasaga 1989). The latest version is GEOCARB II (Berner 1994, 1995a; see figure 3-3). Innovations included mathematical simplification, incorporation of variation of solar luminosity (which has changed about 6% over Phanerozoic time), different assumed feedbacks for carbonate and silicate weathering, updating based on new data on subduction and seafloor spreading over the past 150 million years, changes in the assumed activation energy of chemical weathering of silicates and global runoff, and the inclusion of explicit biological effects on weathering rates (i.e., fertilization of land plants by atmospheric pCO_2 levels, difference in weathering from angiosperms and gymnosperm cover, etc.; more on this in chapter 4).

Phanerozoic Atmospheric Carbon Dioxide Levels

Of particular interest is the predicted variation of atmospheric pCO_2 levels and corresponding global surface temperatures. Estimates of paleoatmo-

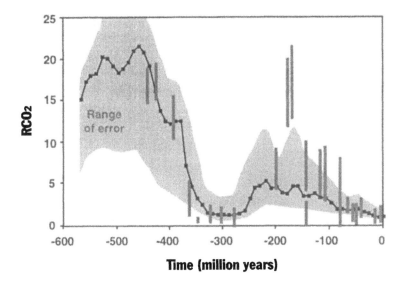

FIGURE 3-4.

Atmospheric carbon dioxide level versus time for the Phanerozoic. RCO_2 is defined as the ratio of the mass of CO_2 in the atmosphere at some time in the past to that at present (preindustrial value of 300 parts per million). The heavier line joining small squares represents the best estimate from GEOCARB II modeling, updated to have the effect of land plants on weathering introduced 380 to 350 million years ago. The shaded area encloses the approximate range of error of the modeling based on sensitivity analysis. Vertical bars represent independent estimates of CO_2 level based on the study of paleosols (after Berner 1997).

spheric pCO_2 levels during the Phanerozoic are now available using several approaches (Yapp and Poths 1996; Mora et al. 1996; McElwain and Chaloner 1995). Yapp and Poths' (1996) estimates are based on the measured level of solid solution carbonate anion in goethite, along with the isotopic composition of its carbon. Mora et al.'s (1996) estimates are derived from the isotopic composition of carbon in paleosol carbonate and organic matter. Estimates of paleoatmospheric pCO_2 levels by McElwain and Chaloner (1995) come from the observed variation of the number of stomata in fossil leaves; modern leaves have fewer stomata after being grown in higher pCO_2 atmospheres. These estimates generally agree well with those computed from Berner's models (figure 3-4). This agreement stands as strong support for the program of making the models of the carbonate–silicate biogeochemical cycle more concrete, a program that can yield empirically testable results and ultimately an explanation of climatic variation over geologic time. The main problem in going from Phanerozoic to Precambrian models is the exponential decrease in the preserved sedimentary record and its chances for postdepositional alteration with increasing age. Thus, Precambrian modeling is more problematic, but more challenging nonetheless.

Introduction to weathering and soil formation; why the typical textbook approach is misleading by compartmentalizing physical, chemical, and biologic weathering (their synergistic interrelations). Biotic enhancement of weathering. The classic studies of Polynov (role of primitive organisms in soil formation in mountainous terrains and succession), Krumbein, Jackson and Keller (lichen weathering on Hawaiian lava flows). Summary of processes contributing to biotic enhancement of weathering (soil stabilization, role of carbonic and organic acids, biophysical weathering, the biota as a sink for calcium, magnesium, regional and global factors including evapotranspiration, frost wedging, oxidation). Tectonics and its influence on climate and weathering (impact of Raymo's ideas on the global climatic impact of the uplift of the Himalayan Plateau; geomorphology and weathering rates; rock chemistry, tectonics, and the carbon sink).

Introduction to Weathering and Soil Formation

We will not attempt here to duplicate the comprehensive reviews of chemical and physical weathering published elsewhere (Berner and Berner 1987; Drever 1988; White and Brantley 1995). Before systematically considering the weathering process, first an observation on its usual treatment. For the most part, geology textbooks treat the process of weathering as either physical (mechanical) or chemical. Biological weathering is either not mentioned or discussion is limited to a couple of examples (e.g., root wedging). A few notable exceptions: Arthur Holmes' classic *Principles of Physical Geology* (latest edition, Holmes 1978) and Russian geology texts (e.g., Gorshkov and Yakushova 1972). The synergism of the three processes in the real world is not discussed at all, or it is limited to mention of the effect of surface area increase as a result of physical weathering on the acceleration of chemical weathering. An extensive body of literature now documents the multifold role of the biota, particularly microbes, in physical and chemical weathering (Yatsu 1988; Banfield and Nealson 1997). For a useful review of the role of plants in mineral weathering see Kelly et al. (1998).

Physical weathering is defined as the breakup of underlying rock into smaller pieces, thus increasing the potential reactive surface area of the component minerals. Abiotic processes (or mainly so) include frost wedging (the cracking of rocks from the expansion of water upon freezing), jointing (cracking of bedrock upon unroofing, the removal of overlying layers by erosion), salt formation, and temperature fluctuations, particularly in the presence of traces of water, as well as cracking (exfoliation, the peeling off of layers like an onion) as a result of chemical weathering. The latter usually results from expansion upon secondary mineral formation such as clay. Frost wedging is usually cited as the most important agent of physical weathering by geomorphologists.

Biogeophysical weathering is complementary to the above processes. The latter includes root wedging, cracking of rock and mineral grains from expansion of fungal hyphae and microbial polysaccharides upon the absorption of water, and ice nucleation associated with lichens (Kieft 1988), as well as the multifold activity of animals burrowing, overturning and mixing the organic and inorganic components of the weathering environment (i.e., the soil). Soil fauna includes earthworms (Darwin 1881, reprinted 1948; Hartenstein 1986), termites, and ants.

Holldobler and Wilson (1990) cited several studies indicating ants' phenomenal activity, especially in tropical soils, carrying organic matter deep in the soil to depths as great as 6 meters, down to their nests. In the forested soils of New England, ants move as much soil as earthworms, even more in tropical forests. The flux of nutrients carried down by ants promotes the growth of tree roots, thus contributing to a positive feedback between plant productivity and soil turnover.

Chemical weathering entails the decomposition of primary minerals by their reaction with water, carbonic acid, oxygen, organic and inorganic acids, and chelating agents. How much of this decomposition is biogeochemical? First, substantial levels of free oxygen are certainly a biogeochemical product. Photosynthesis and burial of organic carbon has generated and maintained free oxygen in the atmosphere, and the concomitant free oxygen in the weathering microenvironment. Thus, oxidation reactions are biogeochemical. Elevated levels of soil carbon dioxide and, in the presence of water, enhanced levels of carbonic acid, as well as organic acids and chelating agents that complex with cations (e.g., oxalic, humic, and fulvic), are all products of biological activity from root/heterotrophic microbial respiration and decay of organic matter and products of the mycorrhizal community, particularly soil fungi, producers of oxalic acid. Lichenic acids are also clearly biogenic. Inorganic acids that attack minerals are produced by sulfur bacteria (sulfuric acid) and nitrifying bacteria (converting ammonium ion to nitric acid). High sulfuric acid levels in soils are found over rocks with high sulfide contents and as a result of mining (Berner and Berner 1987). Nitric acid is generally a minor component of soil acid, but can be significant where anthropogenic sources are involved (acid rain; see Likens and Bormann 1995). These acids, along with the "major players" carbonic and the organic acids, supply protons, thereby speeding up decomposition of silicates. Many experiments have confirmed this role of microbially produced compounds. Moreover, the development of microcolonies of bacteria and other microbes in crack systems results in retention of water, allowing the production of carbonic acid. Organic acid dissolution of silicate minerals and chelation of iron and aluminum are likely particularly important in tropical climates, with their very high net productivities (Thomas 1994).

The microenvironments in soils in contact with mycorrhizae and plant roots (the rhizosphere) and microbial biofilms surrounding mineral grains can have significantly higher organic acid and chelating agent concentrations and lower pH than the soil waters commonly sampled (Berner 1995a). Mycorrhizal fungi attached to silicate mineral grains have been shown to promote chemical weathering (Jongmans et al. 1997). Consider the implications of the estimated surface areas of fungal hyphae, plant roots, and bacteria being 6, 35, and 200 times, respectively, the area of the Earth (Volk 1998). Thus, these biologically created microenvironments probably play an important role in enhancing weathering rates of soil minerals (Robert and Berthelein 1986; Hiebert and Bennett 1992; Barker et al. 1997; Gobran et al. 1998). In their review of the role of organic acids in mineral weathering, Drever and Vance (1994) concluded that organic acids probably have significant impact in accelerating dissolution of mafic minerals in soil microenvironments, where much higher organic acid concentrations are reached than in bulk soil solutions. This conclusion has important implications to enhancing the silicate weathering carbon sink because mafic rocks (e.g., basalt) contribute disproportionately to this sink owing to their high levels of CaMg silicates compared with more silica-rich rocks such as granite.

Although the case for the overall effect of biota on chemical weathering is strong, biotic effects also may locally and temporarily reduce the rate of chemical weathering. The latter include development of macropores in soil (e.g., from decayed roots), which allow most of the incoming water to bypass reactive mineral grains, and coatings on rock and mineral surfaces of micro- bial origin, such as rock varnish formed in arid regions (Krumbein and Jens 1981), which protects mineral grains from chemical decomposition.

Nevertheless, the synergistic interaction of all these weathering processes creates a biogeochemical and biogeophysical interface between the atmo- sphere and crust (mainly continental, but also involving volcanic islands), namely, soil. The extensive literature on the differentiation of soil and the key role of climate (especially precipitation and temperature) in determining the rich variety of soil types will not be reviewed here. A simplified picture of the differentiation of a modern soil is shown in figure 4-1. The O zone is composed largely of leaf litter and humus; just below is zone A, where leach- ing has taken place, followed by zone B, the zone of accumulation. In tem- perate climates, zone B is typically enriched in iron oxides and clay; in arid climates, calcium carbonate (caliche) accumulates. Finally, zone C is the transition to fresh bedrock below. In zone C the bedrock is partially disinte- grated (physical weathering) and decomposed (chemical weathering). There are many variations to this basic pattern. In situ soils progressively thicken and differentiate over time, as contrasted to transported soil such as clay de- posited on a flood plain, which may be deposited and then eroded by higher river discharge.

This idealized pattern of vertical stratification is often disrupted by biotur- bation as a result of root decay and collapse, and burrowing animals ranging from mammals to ants (see Brimhall et al. 1991; Butler 1995). With these processes, along with the flux from root activity, organic matter is thereby carried deep into zone C, where biogeochemical and biogeophysical weath- ering may be accelerated. Richter and Markewitz (1995) emphasize the im- portance of the zone C horizon to weathering, drawing on their studies of a deep, clay-rich soil in South Carolina. They measured progressively increas-

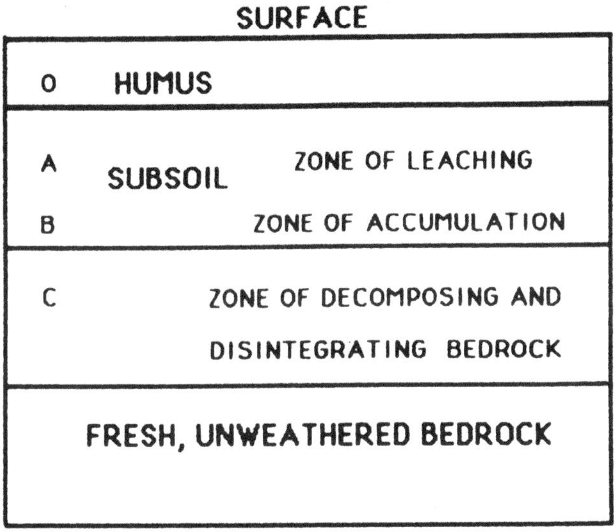

SURFACE

O	**HUMUS**	
A	**SUBSOIL**	ZONE OF LEACHING
B		ZONE OF ACCUMULATION
C		ZONE OF DECOMPOSING AND DISINTEGRATING BEDROCK
	FRESH, UNWEATHERED BEDROCK	

FIGURE 4-I.
A typical soil profile.

ing carbon dioxide levels in soil air with depth down to 6 meters. These high carbon dioxide levels, apparently generated mainly from rhizosphere activity, are a major source of chemical weathering via carbonic acid production.

The multifold processes occurring in soils that link the physical/biological/chemical constitute a complex whole in the real world. Although a reductionist program has had great explanatory power in elucidating weathering as a function of time and space, a fuller understanding must come from a comprehensive systems approach that illuminates the rich interactions, linear and nonlinear, that generate a dynamic soil. This is a research program that brings together diverse disciplines in geology, biology, physics, chemistry, and even astronomy.

Russian researchers such as Polynov (1945, 1948, 1953; and his pupils such as Yarilova 1950; a fascinating account is found in Lapo 1982), following Vernadsky, long ago held that chemical weathering and soil formation on the Earth's surface is largely determined by biological activity. Their studies in mountainous areas of the Soviet Union (e.g., Pamirs) revealed a regular pattern of ecologic succession, starting with the first colonization of exposed plutonic and metamorphic rock by primitive organisms.

The modern succession of life upon colonization of bare exposed rock is, as a first approximation, similar to the historical biotic succession on land ("ontogeny recapitulates phylogeny" here in the same metaphorical sense as for the development of a single organism). The first colonizers of bare rock are typically cyanobacteria, along with actinobacteria and microfungi (heterotrophs). These are soon followed by photosynthesizing green algae and then crustose lichens, the variety that is closely attached to the rock surface. There is now evidence for a nearly contemporaneous emergence of vascular plants and lichens in a modern sense (i.e., with true fungi, mainly *Ascomycetes* in the early Paleozoic); primitive lichen-like symbioses such as actinolichens may have emerged in the Precambrian (see further discussion in chapter 8). Mosses and then vascular plants commonly occupy the crevices and crannies of the rock outcrop (often simultaneously with the adjacent lichens), with a progressive occupation of the rock surface to follow, crowding out the lichen community. Lichens generate microsoil, which ultimately contributes to a primary soil upon which the vascular plants thrive. The slope of rock outcrop has a strong influence on the speed and completeness of the succession outlined here. For high slopes, bare rock remains largely soil free, with succession stopped at the lichen community, consisting of crustose, foliose (leaflike), and fructicose (branching) varieties, as well as mosses, with higher plants limited to the crevices. On shallow slopes, the succession rapidly goes to the higher plant-dominated ecosystem, particularly in temperate and tropical climates. Higher plants also may be the pioneers, particularly on porous volcanics, when seeds and traces of organic matter are blown into crevices and vesicles and sprout before even lichens get a foothold.

Krumbein's studies on microbial degradation of rock, starting in the 1960s, also led him to the conclusion that weathering is largely dominated by biological activity and to a geophysiological theory of the Earth's surface activity (Krumbein and Dyer 1985; Krumbein and Schellnhuber 1990, 1992; Krumbein and Lapo 1996). Another strong proponent of the role of microbes in accelerating weathering has been Eckhardt (1985).

Lichen Weathering

One organism's weathering potential has been studied for over 100 years: lichens, a symbiosis of fungi (the mycobiont) with a phototrophic partner (the photobiont), algae or blue green algae, more precisely cyanobacteria. For two recent surveys of lichen biology and ecology see Ahmadjian (1993) and Nash (1996). Early work focused on the identification of calcium oxalate, derived from oxalic acid reactions with calcium-bearing substrates, as well as the description of hyphae breaking up and partially decomposing silicate and carbonate mineral grains (e.g., Bachmann 1904; Stahlecker 1906; Fry 1924 1927). The role of lichens in weathering of exposed rock is well established (see reviews by Syers and Iskandar 1973; Jones and Wilson 1985; Wilson 1995). Lichens apparently accelerate physical and chemical weathering of their substrate by rhizine penetration with thallus expansion and contraction, and by the production of carbonic and organic acids (e.g., oxalic) and chelating agents. Lichen weathering effects include etching of substrate grains and synthesis of secondary minerals such as iron oxides, silica, and possibly clay (see reviews by Jones 1988; Wilson and Jones 1983; Wilson 1995) (figure 4-2).

A pioneering study compared weathering rates of lichen-covered and lichen-free basalt lava flows of recent age in Hawaii (Jackson and Keller 1970a). Lichens appeared to have increased weathering rates by one to two orders of magnitude compared with adjacent bare rock, based on measurements of relative weathering rind thickness and chemistry. We will return to a discussion of this study in chapter 5, in our consideration of estimates of the biotic enhancement of weathering.

The Biotic "Sink" Effect and Soil Stabilization

In addition to the effects discussed above, other factors contributing to the biotic enhancement of weathering include the biotic sink effect and, probably the single most important effect, the biotic stabilization of soil.

The biota is itself a sink for soluble products of weathering, including potassium, calcium, and silicon; thus, according to Le Chatelier's principle, weathering reactions decomposing silicates are enhanced because lowering the concentration of the products tends to drive the reaction forward. Microbes and many higher plants are accumulators of these products (e.g., fungi accumulate potassium, and horsetail [Equisetum] and most plants [especially grasses] accumulate silica, which serves as a structural component). Cations are then made available again to soil solutions upon decay of the organic

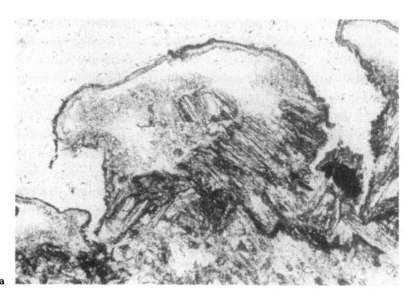

a

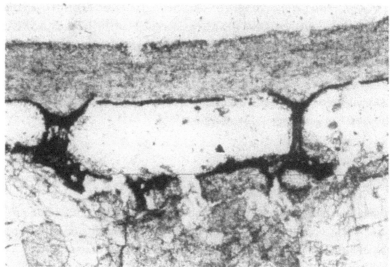

b

FIGURE 4-2.

Lichen-induced weathering of silicate rocks (photomicrographs of thin sections). **a:** Crustose lichen (*Aspicilia* species) on Soldiers Delight, Maryland serpentinite. Weathering penetrates to 200 to 400 microns below the rock surface. Iron oxide formation and depletion of magnesium occurs in the weathering zone. **b:** Foliose lichen (*Flavoparmelia baltimorensis*) on Herndon, Virginia basalt. Thallus attached by rhizones (*black*). Altered plagioclase and pyroxene, iron oxides in weathered rind several hundred microns thick.

matter, with a fraction remaining adsorbed and bound to the organics and clay in the soil and subsoil. If the soil mass and biomass are in steady state, there must be erosion of soil containing these weathering products balancing new soil development. This effect is of course magnified on sloping terrain, where tree falls, severe storms, fires, and disease all tend to episodically increase soil erosion. Hence, in relation to the actively decomposing silicates, under these conditions, or especially in the case of actively aggrading biomass as for rapidly growing plant cover, the latter is a net sink for soluble products and must be included in weathering budgets to derive realistic mineral weathering rates in soils (see discussion in chapter 6). In the case of rapid turnover of biomass, as in a mature rain forest with very low rates of soil erosion, these effects make a small contribution to the flux out of the ecosystem (Alexandre et al. 1997).

The most important single role of biota in accelerating the rate of chemical weathering is probably the stabilization of soil. A soil, particularly in the zone of active decomposition and disintegration of bedrock, greatly increases the chemical weathering of silicates for a given land area because of the high surface area of the mineral grains relative to a bare rock surface. For example, a mineral soil 1 meter thick with individual grains (cubes) 1 mm^3 in volume has 6000 times the potential reactive surface area of a bare rock surface with no permeability. In a real soil, much of the potential surface area is blocked off by coatings or bypassed by channels (macropores) that function as conduits for water. An additional property of soils that significantly contributes to enhancing chemical weathering is the sponge effect, the retention of water available to react with minerals. Water is retained in a soil by surface tension and soaking up by the organic and living components, slow evaporation, and runoff compared with bare rock.

The key role of vegetation in impeding soil erosion is well known from studies around the world (dramatically illustrated in figure 4-3). In 1876, Frederick Engels vividly described the effects of deforestation:

What did it matter to the Spanish planters in Cuba, who burned down forests on the slopes of the mountains and obtained from the ashes sufficient fertilizer for one generation of very highly profitable coffee trees . . . that the heavy tropical rainfall afterwards washed away the now unprotected upper stratum of the soil, leaving behind only bare rock? (1940, pp. 295–296).

FIGURE 4-3.
A blackjack oak in Tennessee, perched on a soil pedestal surrounded by eroded ground, illustrating soil stabilization by plants. (Courtesy of U.S. Forest Service.)

The key role of vegetation in impeding soil erosion is well known from studies around the world. Higher plants stabilize soil with their subsurface root system and by the creation of organic litter on the ground, all of which slow down erosion. In addition, biotic products such as mucilaginous poly-saccharides effectively bind soil particles. The lowest rates of soil erosion occur today in forested land, typically 10^{-2} to 10^{-4} times the rates of bare,

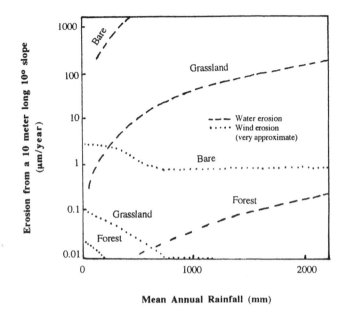

FIGURE 4-4.
Soil erosion rates as a function of rainfall and vegetation regimes. (After Lal 1990.)

unvegetated ground (Zachar 1982) (figure 4-4) for soil erosion rates as a function of rainfall regimens and vegetation). Figure 4-4 illustrates a summary of data from many empirical studies of soil erosion from both water and wind on a variety of vegetated and bare surfaces. Note that bare surfaces have consistently the highest erosion rates, forests the lowest. As rainfall increases, there is a saturation effect on water erosion. On the other hand, high rainfall decreases wind erosion as a result of the binding effect of wetting soil, again with a saturation effect.

The stabilization effect is not limited to higher plants; microbial (cryptogamic) soils found in desert regions and in the early stages of colonization of exposed rock surfaces are stabilized by algae and lichens (Campbell 1979; Thornes 1990) (figure 4-5). Indeed, the use of off-road vehicles in American deserts breaking up the microbial crust has led to rapid soil erosion. One study found an exponential decline in splash erosion (resulting from impacting raindrops) as cryptogam cover increased (Eldridge and Greene 1994). Treub (1888; cited in Thornton 1996) noted that 3 years after the catastrophic eruption of Krakatau in 1883, six species of what we now recognize

FIGURE 4-5.
Microbial soil: lichen on basalt flow, Hawaii. (Courtesy of Ford Cochran.)

as cyanobacteria grew as a gelatinous mass on the newly formed volcanic ash substrate. Modern day microbial soils may be models for Precambrian soils, before the emergence of land plants (Campbell 1979; Retallack 1990).

A sterile land surface for several billion years in the Precambrian is implausible given that procaryotes stabilize soils even today. Fossil soils (paleosols) of the Precambrian age show indications of a contemporary surface biota. These include the presence of reduced carbon, trace element patterns reminiscent of biotic influence, and features also found in some modern soils (e.g., blocky structures indicating coatings of microbial polysaccharides) (Retallack 1990). The existence of Archean and Proterozoic paleosols with thicknesses up to 80 meters (Miller et al. 1992; Marmo 1992) is also consistent with a contemporary stabilizing microbial cover. More direct evidence from microfossils found in chert from paleokarst deposits at 0.8 and 1.2 Ga (Horodyski and Knauth 1994) supports the existence of a microbial land biota (in this case probably procaryotes, as identified in thin section). The latter conclusion is supported by carbon and oxygen isotopic signatures consistent with subaerial biotic fractionation from paleosols at 0.8 to 1.2 Ga (Beeunas and Knauth 1985; Horodyski and Knauth 1994). The isotopic ratio of carbon from primary dolomites (CaMg carbonate) is most consistent with an isotopically light carbon input from meteoric water flushing through a biologically active soil contributing respired carbon dioxide. A 2.0 to 2.2 Ga paleosol from South Africa has been interpreted as a laterite with relict soil textures, suggesting development under a biotic cover accompanied by hot and humid conditions (Gutzmer and Beukes 1998). In particular, this paleosol has a leached upper zone consistent with organically complexed iron being transported to deeper levels. If these inferences are correct and a land biota first appeared in the Archean, chemical weathering rates should have been higher as a result of their soil-stabilizing effect.

In addition, physical weathering processes are likely to be different on an abiotic rock surface. Frost wedging, the most important contributor to physical weathering, may be much more limited or even absent if abiotic temperatures are significantly higher than biotic temperatures. On the other hand, higher surface temperatures with large diurnal fluctuations could promote thermal cracking, as seen in deserts today. On an abiotic land surface, water and wind erosion are likely to preempt significant accumulation of soil, even on slight slopes. Mineral grains freed up by weathering should expected to be rapidly transported by running water and wind relative to biotic conditions, ending up ultimately in ocean basins. The residence time of such sediments in river channels is likely shorter in the abiotic case because of the "flashy" behavior of abiotic streams and rivers, given the lack of water storage in soil (Schumm 1968).

Under biotic conditions, rapid erosion on high slopes, the apparent locale for most contemporary chemical denudation (Stallard 1992), leads to acceleration of chemical weathering because soils reform rapidly with biotic colonization and stabilization. The rate of carbon dioxide consumption by silicate weathering can reach a steady state independent of mineral reactivity under restricted conditions, which include a constant rate of generation of new mineral grains available for reaction (Lasaga and Berner 1998). However, under the dynamic conditions just outlined, this steady state is unlikely to be approached. Further, the rate of generation of new reactive mineral surface area is likely to be much higher in the case of biotically mediated conditions, which include soil stabilization. The contrasting effects of erosion under abiotic and biotic conditions are shown in figure 4-6. Because microbially crusted soils are much more easily eroded than those stabilized by plants, the evolution of the land biota has plausibly resulted in a progressive increase in the biotic enhancement of weathering. Note that contemporary procaryotic desert crusts are absent on steeply sloping terrain (Campbell 1979).

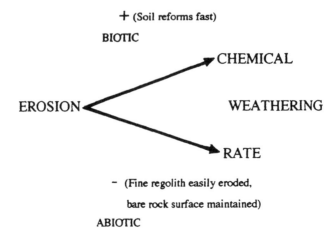

FIGURE 4-6.
The effect of erosion on chemical weathering rate under biotic and abiotic conditions.

Regional and Global Biotic Effects

Except for the role of atmospheric oxygen, so far we have focused on direct and local biotically mediated effects. On the larger scale of regions and the globe, other phenomena occur that probably act to increase the overall biotic enhancement of weathering.

The diversity of effects on different scales entailed by biotic enhancement of weathering is shown in figure 4-7. The local and direct effects are perhaps better understood than the global effects. The latter may include the effects of global cooling resulting from the evolution of land biota, a topic to be discussed in chapter 8. Thus, frost wedging, a major mechanism of physical weathering, may be biotic in the sense of arising from biotically mediated global cooling, triggering significant ice formation in mountains. The problematic role of atmospheric oxygen includes the possible increase of productivity of land biota by virtue of an ozone shield, to be discussed in more depth later in chapter 8. The emergence of new microbial soil consortia with the rise of atmospheric oxygen in the Proterozoic could have promoted the acceleration of dissolution of CaMgFe silicates by the oxidation of ferrous iron and the organic chelation of its product, ferric iron, thus driving forward the weathering process and intensifying the carbon sink.

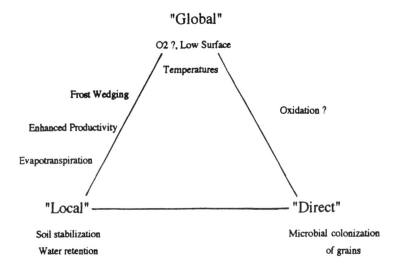

"Global"

O2 ?, Low Surface

Temperatures

Frost Wedging

Oxidation ?

Enhanced Productivity

Evapotranspiration

"Local" ——————————————— **"Direct"**

Soil stabilization   Microbial colonization

Water retention   of grains

Organic acid production, pCO$_2$ elevation in soil,

Stirring up of soil by collapse of channels left by decay of roots and burrowing animals

(Brimhall et al., 1991), biotic sink for Ca, Mg, Fe ?

FIGURE 4-7.

A summary of the influences contributing to the biotic enhancement of weathering.

Another important regional effect is likely, the role of evapotranspiration, as Lovelock (1987) emphasized in his discussion of the geophysiology of Amazonia. Evapotranspiration coupled with precipitation accounts for an efficient recycling of water through the forested ecosystem. In the absence of evapotranspiration from a vegetative cover on land, chemical denudation should be reduced from the lack of water flow-through alone (Berner 1992; Drever 1994). It is interesting that sphagnum moss peatlands also recycle their water efficiently by the same mechanism as forested ecosystems (Klinger 1991). One can therefore speculate that the evapotranspiration/precipitation cycle amplified chemical denudation in terrestrial microbial and bryophyte mat ecosystems in the Precambrian and early Paleozoic.

We will return to the evolutionary aspects of this discussion in chapter 8. Attempts to quantify the combined effects, particularly in relation to the regulation of atmospheric carbon dioxide levels and surface temperature, are discussed in the next chapter.

Tectonics and Its Influence on Climate and Weathering

Two prominent hypotheses have been offered on the relationship of tectonics to chemical weathering intensity and long-term climate regulation. Some researchers have presented evidence, mainly from the Amazon and Congo River watersheds, that there is a strong relief effect on weathering rates; high relief is correlative with high denudation rates, whereas the lowest rates occur in the flat lowlands (Stallard 1995a, 1995b; Edmond et al. 1995; Gaillardet et al. 1995). These authors distinguish between weathering and transport-limited erosional regimes. For the former, the rate of transport exceeds the rate of chemical weathering (high slopes), whereas for the latter the reverse is true, so thick soils develop with low denudation rates.

The other main hypothesis is centered on the increase in mechanical erosion and hence chemical erosion driven by uplift, an idea apparently first suggested by T. C. Chamberlain (1899; see Raymo 1991) as an explanation of the initiation of the ice ages. These proposals will be examined more closely in the following discussion, also touching on related issues regarding the interrelation of rock chemistry and tectonics to the carbon sink with respect to the atmosphere/ocean system.

Does Uplift of Mountains Promote Chemical Weathering?

In the Stallard model of erosional regimes, "a complex interplay of chemical, physical, and biological processes" generate weathering products (1995a, 1995b). More precisely, these researchers claimed that maximum chemical denudation rates are achieved in the weathering-limited regime, with vegetation maintains a relatively thin soil on steep slopes, trapping water and generating bioacids in contact with fresh rock. With erosion triggered by earthquakes, treefalls, and so forth, fresh rock is reexposed for the regeneration of soil (Stallard 1992). In contrast, in the transport-limited regime, thick highly weathered soils ultimately slow down chemical denudation by shielding fresh rock buried deep below the surface from percolating water. Stallard (1995b) pointed out that in flat terrains vegetation has two contradictory effects on chemical denudation: its stabilization of thick leached soils tends toward its reduction, whereas the production of bioacids and penetrating root systems tends toward its increase. In any case the estimated chemical denudation rates even in flat terrains with thick soils developed on granitic rocks are still some two orders of magnitude higher than a computed two-dimensional bare rock case. It is interesting to speculate that this mechanism could explain at least part of the cooling apparently tied to global periods of orogenesis (e.g., possibly at 2.5 Ga; see Knauth and Clemens 1995). More mountains could magnify biotic influence on weathering if the Stallard model is correct.

Estimated chemical denudation rates of siliceous rocks from the humid tropics of South America support the Stallard model (figure 4-8). More recent data from the Guayana shield (Edmond et al. 1995) and Congo basin (Gaillardet et al. 1995) provide additional support for this scenario. In both cases, significantly lower chemical denudation rates were inferred from rivers draining relatively flat terrain. In contrast, White and Blum (1995) found no correlation between chemical denudation rates of granitic rocks and relief in their study of 68 watersheds. They concede that their data set contains only one tropical watershed (rain forest in Puerto Rico). Drever and Clow (1995) suggest that White and Blum's negative finding with respect to the influence of relief may have resulted from the absence of really low relief areas (e.g., Amazon basin), or that "noise" in data prevented the emergence of a correlation. The basic validity of the Stallard model will be assumed in the discussion of the evolutionary history of the biotic enhancement of weathering in chapter 8.

An even more radical view of the factors influencing weathering has emerged, namely that climate has very little to do with determining the rates of chemical denudation and carbon dioxide consumption by silicate weathering, but rather the predominant controls are relief and lithology (Edmond and Huh 1997; Huh and Edmond 1997; Allegre et al. 1997; Gaillardet et al. 1997). This contention is based on inferred carbon dioxide consumption rates from a global data base of river and watershed rock chemistry. Silicate and carbonate weathering contributions to rivers are distinguished by using the different strontium isotopic signatures of each lithology.

However, supporters of a strong climatic effect on weathering rates have noted that the carbonate weathering contribution is often underestimated in watersheds with predominant silicate lithology (Berner and Berner 1997). For example, significant outcrops of marble occur in the cold Aldan River watershed in Siberia (Berner and Berner 1997), which Huh and Edmond

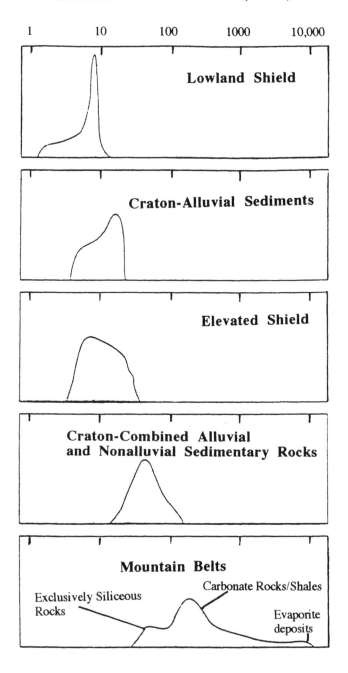

Estimated denudation rate (m/Ma)

Lowland Shield

Craton-Alluvial Sediments

Elevated Shield

Craton-Combined Alluvial
and Nonalluvial Sedimentary Rocks

Mountain Belts

Carbonate Rocks/Shales

Exclusively Siliceous
Rocks

Evaporite
deposits

FIGURE 4-8.
Histogram of denudation rates for morphotectonic regions in the humid tropics of
South America. (After Stallard 1992.)

(1997) claim has a silicate denudation rate comparable with that of the tropics. Furthermore, the strontium isotope ratio ($^{87}Sr/^{86}Sr$) of carbonates may overlap with silicate lithologies. In addition, experimental weathering of granite-like rocks indicates that the strontium isotope ratio released to water changes with time; at first ubiquitous trace calcite weathers out, giving a low ratio, then radiogenic biotite, then less radiogenic plagioclase (White et al. 1997). In view of this behavior, the researchers noted that strontium isotopic ratios should be used with caution as signatures of weathering because freshly exposed silicate rocks could yield soluble strontium on weathering with significantly different isotope ratios than depleted already weathered rocks.

Evidence for climate control (temperature, runoff, and implicitly vegetation) includes experimental demonstration of the temperature effect on mineral dissolution rates, patterns of clay mineralogy in soils indicating progressive increase in silicate weathering with increasing temperature (higher kaolinite and gibbsite) (Berner and Berner 1997), and the evidence for biotic enhancement, to be discussed further in the next two chapters. When one lithology is looked at as in White and Blum's (1995) study of granitic watersheds, a significant influence of climate emerges. Another lithology, basalts from four volcanic islands, gives a strong correlation between chemical denudation rates, calcium and magnesium in particular, and temperature (Louvat 1997).

Raymo's Hypothesis

In an influential series of papers, Maureen Raymo has argued that Cenozoic global cooling resulted from the uplift of the Himalayas (Raymo et al. 1988; Raymo 1991; Raymo and Ruddiman 1992; Edmond 1992). The proposed mechanism entails the increase of physical erosion and precipitation (monsoonal rains) as a result of uplift, which in turn promote chemical weathering of silicate rock and the increase in carbon sink. The documented rise in $^{87}Sr/^{86}Sr$ ratio of seawater since the Cretaceous as measured in marine limestones is cited as support for this model, because it is claimed that such a rise could only result from an increase in radiogenic strontium derived from continental weathering (high Rb/Sr ratio and age compared with oceanic basalt) (figure 4-9). Furthermore, glacial erosion resulting from Himalayan uplift grinding up rock particles is claimed to have significantly increased silicate weathering (Sharp et al. 1995).

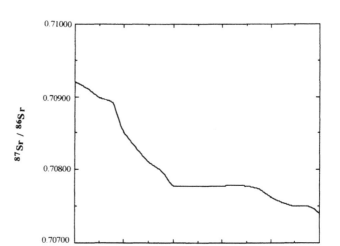

FIGURE 4-9.

Strontium isotope ratio as a function of age for ocean, derived from measurements on deep-sea carbonates. (Smoothed curve, after Edmond 1992.)

All of the preceding arguments have now been cogently challenged. First, it is not clear that the trend in $^{87}Sr/^{86}Sr$ ratio of the ocean in the last 100 million years is tracking an increase in continental weathering or a change in the continental source material to more radiogenic strontium (Berner 1995a). Derry and France-Lanord (1995) found that in the mid-Pliocene with reduced erosion of the Himalayas, the $^{87}Sr/^{86}Sr$ ratio of the ocean increased. Furthermore, most of the rock particles transported to the Bengal fan are not weathered chemically. Harris (1995) has explained much of the rise in the $^{87}Sr/^{86}Sr$ ratio as a result of weathering of radiogenic metasedimentary rocks exposed during uplift. Blum et al. (1998) found that the strontium flux and high $^{87}Sr/^{86}Sr$ ratio in Himalayan rivers is likely derived from trace amounts of calcite in silicate rocks and not from silicate weathering itself (only the latter is a net consumer of carbon dioxide from the atmosphere). Godderis and Francois (1995) argued from their modeling of the strontium and carbon cycles that Cenozoic cooling resulted mainly from the decrease in outgassing at the mid-oceanic ridges and changes in the chemical weathering rates in the rest of the world excluding the Himalayas.

Second, the timing of the uplift of the Tibetan plateau apparently does not coincide with the climate history (Caldeira and Searle 1997). The plateau's elevation was higher 8 million years ago than today; therefore, cooling in the last 8 million years, including the glacial episodes of the Pleistocene, cannot be simply explained by its uplift.

Evidence for enhanced silicate weathering from glacial erosion is also equivocal because it has not been proven that carbonates and ion exchange reactions do not dominate the measured fluxes of Ca^{2+} and HCO_3^- (Berner 1995a). White and Blum (1995) found no correlation between chemical denudation in granitic terrains and the presence or absence of prior glaciation. Looking at the available data from some ten different glacier-covered catchments, Anderson et al. (1997) found that glacial water is enriched in potassium and calcium relative to other cations, but silica is low. They attribute the elevated calcium to dissolution of trace carbonate (not a carbon sink) and cation leaching from biotite (a potential sink if magnesium uptake by the biotite is less than calcium loss). However, they concluded that silicate weathering is inhibited by low temperatures and lack of vegetation. The grinding of minerals apparently produced a temporary (few years?) enhancement of dissolution that is analogous to the artifact effect produced by grinding powders in the laboratory (see discussion in chapter 5). This short-lived enhancement also may play a role in the continued exposure of reactive mineral surface by ice-induced cracking of rock in environments that are above freezing in the day, but below freezing at night (e.g., mountains, summers in Arctic and Antarctic). This effect may be at least partially responsible for the enhanced silicate weathering rates inferred along an Antarctic stream (Blum et al. 1997). Furthermore, although the authors believe weathering in this locale is abiotic, it is plausible that the presence of a microbial biota contributes to enhanced mineral dissolution in the hyporheic zone (sediment bordering stream), where the chemical weathering is argued to take place, given that liquid water in cold environments is hardly sterile. A recent study of glacial environments demonstrated the apparent presence of microbially mediated weathering (Skidmore et al. 1997).

A major problem with Raymo's hypothesis is that unless a concomitant source of carbon dioxide to the atmosphere is present that balances the proposed increase in the silicate weathering sink, the whole inventory of carbon dioxide in the atmosphere/ocean pool will be removed in a very short time (Volk 1993; Caldeira et al. 1993; Berner and Caldeira 1997). For example, a 10% excess of the sink today would remove all the carbon dioxide from this pool in 5 million years (Volk 1993). There is no evidence for such drastic reduction of carbon dioxide in the Cenozoic. To address this issue, Raymo (1994) proposed that a decrease in organic matter burial (an equivalent carbon dioxide source to the atmosphere/ocean) kept the atmospheric carbon dioxide level from declining. However, this proposed solution has its own problems; there must be near perfect variation in organic matter burial to keep the carbon dioxide level in the atmosphere in balance (Volk 1993), but unlike silicate weathering, there is no evidence of a feedback between organic matter burial and the level of atmospheric carbon dioxide (Berner and Caldeira 1997; Broecker and Sanyal 1998).

One other mechanism for maintaining a steady-state level of carbon dioxide has been proposed; it is regulated by aerial volcanism, which increases both the rates of release of carbon dioxide to the atmosphere and its removal by silicate weathering of the more easily weathered volcanics, especially with higher levels of carbon dioxide in rain (Allegre et al. 1997). However, weathering under these conditions is likely to be strongly mediated by climate (see Louvat 1997), and there is little support for a strong component of volcanic weathering in global fluxes.

Finally, in an interesting twist to this story, Molnar and England (1990) have argued that late Cenozoic global climatic change itself led to the continued uplift of the Tibetan Plateau. Global cooling would have increased latitudinal temperature differences, perhaps facilitating the monsoon and the increase in erosion rates and isostatic rebound. They suggest that climate change, weathering, erosion and isostatic rebound might be linked in a positive feedback, thereby enhancing a negative greenhouse effect. However, the carbonate-silicate stabilizer will kick in at time scales of 5×10^5 to 10^6 years, limiting any negative greenhouse effect.

On the other hand, Brozovic et al. (1997) argued the opposite influence of climate on uplift as proposed by Molnar and England (1990), that is, that warmer or drier climates raise peak elevations of mountains by elevating the erosive action of glaciers. They noted that both maximum and mean mountain elevations presently decrease with increasing latitude.

Rock Chemistry, Tectonics, and the Carbon Sink

Another recent theory regarding the relationship of tectonics to climate deserves mention. Reusch and Maasch (1995) proposed that the uplift and weathering of magmatic arcs rich in mafic rocks have intensified the carbon sink with respect to the atmosphere. This enhancement is plausible because mafic rocks are richer in CaMg silicates, particularly those that react fastest with soil acids (e.g., olivine, calcium-rich plagioclase). A prominent example in the Cenozoic is New Guinea, with abundant ophiolites, situated in a tropical weathering regime. They estimate an enhancement of 8.6 times in the weathering rate of simatic crust (mafic and ultramafic rocks rich in calcium and magnesium) relative to sialic crust (rich in silicon) (Reusch and Maasch 1995). However, one study on chemical weathering rates in tropical areas has indicated a more modest enhancement for the case of basalt relative to acidic volcanic rocks (Amiotte Suchet and Probst 1993). Reusch and Maasch's hypothesis is currently under discussion. The Cenozoic record of oxygen, strontium, and osmium isotopes does not appear to support their proposal for the role of New Guinea (Caldeira, personal communication). Nevertheless, the role of rock chemistry in the codetermination of the carbon sink along with climate and biota is likely important and merits further study.

Estimates of biotic enhancement of weathering from field studies (alpine terrains, lichens, and experimental biotic weathering). The laboratory/field rate paradox: is biology irrelevant? (experimental studies of mineral dissolution; can observed chemical denudation rates in watersheds be predicted from experimental/theoretical models? why do most models based on laboratory rates of dissolution predict rates one to three orders of magnitude higher than observed field rates? possible role of subsurface piping, saturation effects).

This chapter will focus first on the several field and laboratory studies attempting to infer an estimate of a biotic enhancement of weathering on the contemporary Earth, then examine the paradox of why most models based on laboratory data on silicate dissolution rates (presumed to be abiotic) predict field rates that are one to five orders of magnitude higher than those observed in studies of chemical denudation in soils and watersheds. We will conclude with my suggested decomposition of the combined effects producing the observed field rates, with a goal of possibly resolving the above paradox.

Estimates of Biotic Enhancement of Weathering from Field Studies

ICELAND

The earliest study attempting to estimate the effect of vegetation on chemical weathering rates was apparently that of Cawley et al. (1969). Their estimate of biotic enhancement was about 3, based on a comparison of bicarbonate concentrations (not fluxes, which would be expressed for example as moles km^{-2} yr^{-1}) in water draining barren and vegetated basalt lava in Iceland. They concluded that plants have had only modest effects on chemical weathering. Several caveats are in order in the interpretation of the results of this pioneering study that attempted to quantify the biotic role in weathering. First, it is not clear whether the bicarbonate is derived from weathering of minor carbonate phases in the rock (if so, it would not represent a carbon sink because the same number of moles of carbon dioxide consumed in weathering is released on calcium carbonate precipitation in the ocean) or of silicates. Second, the proper basis for comparison of the two terrains, the actual fluxes, were not measured or derived. Finally, the so-called barren area is almost certainly colonized by bacteria, algae, and lichens (Thomson 1984), organisms that contribute in some degree to chemical weathering. Even cyanobacteria apparently solubilize basaltic glass in Iceland (Thorseth et al. 1992). A follow-up study in Iceland comparing bare, partly moss- and lichen-covered surfaces with tree-covered soil gives a weathering enhancement in calcium and magnesium by the latter soil of two to five times (Moulton and Berner 1998).

Gislason et al. (1996) found in their study of chemical weathering in Iceland that fluxes of calcium and magnesium increase with vegetative cover at constant runoff. Here vegetative cover, determined by remote sensing techniques, consists mainly of higher plants. Mosses and lichens are only partially captured in their estimates of vegetative cover, and microbial growth not at all (Gislason, personal communication). Thus, they actually determined the biotic enhancement of higher plants over a mixed bryophyte/lichen/microbial cover for this particular locale (with an average temperature of only 5°C). Their data indicate roughly an order of magnitude enhancement as defined above. They conclude that the main effect of vegetation on chemical weathering in their locale is the lowering of soil pH, destabilizing secondary minerals containing calcium and magnesium that would be stable at higher pH.

LICHEN WEATHERING

The other, now classic study attempting to estimate a biotic enhancement of weathering is that of Jackson and Keller (1970a), a study introduced in chapter 4. Despite a fairly extensive body of literature on lichen weathering, dating back to the late 19th century, their paper remains the only explicit estimate of a lichen chemical weathering rate (we published to date an abstract:

Cochran and Schwartzman 1995). An estimate of the lichen enhancement of weathering can be inferred from measurements of weathered rind thickness (and chemistry) of bare and lichenized basalt flows. Their measurements give an inferred local lichen chemical denudation rate of at least 10 to 100 times the rate on bare rock, which is less than 3×10^{-5} mm yr^{-1} (figure 5-1).

Thus, taken at face value, the local lichen enhancement is about one order of magnitude lower than typical chemical denudation rates in tropical/saprolitic soils (about 0.05 mm yr^{-1}). There is now some controversy over the interpretation of their results; see Berner (1992), with comments by Jackson (1993) and Schwartzman (1993), with reply by Cochran and Berner (1993a). Berner (1992) and Cochran and Berner (1993a) did not deny that lichens chemically weather their rock substrate, but argued that the *Stereocaulom vulcani* colonies they studied in Hawaii did not show such weathering. In their reply, Cochran and Berner conceded that blown in airborne volcanic rock particles captured by the *S. vulcani* thallus were chemically decomposed. Jackson (1993) strongly upheld his original interpretation that the observed iron oxide (ferrihydrite) rind (no clay) under the lichen is a result of weathering of the basalt substrate. Note that hematite and clay, in trace amounts, were only detected in the weathered surface of adjacent bare rock, indicating the biogenic character of ferrihydrite generation under the lichen (Jackson and Keller 1970b). Ferrihydrite also has been reported from the interface of *Pertusaria,* a crustose lichen with basalt (Jones et al. 1980) and as an endolithic lichen weathering product (Johnston and Vestal 1989). Elsewhere, a lichen of the same genus, *S. vesuvianum,* apparently contributes to the chemical/physical breakdown of underlying leucite-bearing basalt, as shown by etching of minerals and x-ray diffraction studies (calcium oxalate was found as a probable component) (Adamo and Violante 1991). Jackson's conclusion that *S. vulcani* chemically weathered its rock substrate was supported by new field and laboratory studies in Hawaii (Brady 1997).

Saxicolous lichens (rock-growing) have several advantages as a natural model of biological weathering. Lichens are really miniature ecosystems. Besides the defining symbionts, they commonly contain a community of microbes, including diatoms and chrysophytes, which we observed in the lower thallus of *Flavoparmelia baltimorensis,* a common saxicolous species (figure 5-2); this finding is significant because of the structural involvement of silica in these accompanying microbes, which apparently act as a sink for silica freed up by weathering of the substrate (Schwartzman 1991). We have sug-

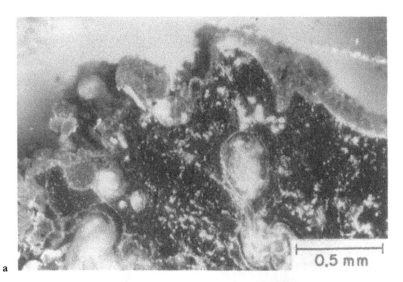

a

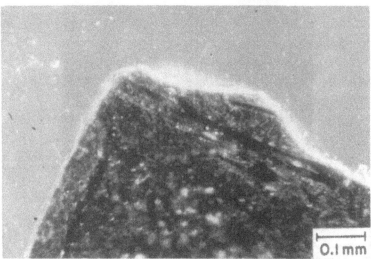

b

FIGURE 5-1.

Hawaiian lichen/rock and bare rock weathering. (Photomicrographs of thin sections, Jackson 1968.) **a:** Weathering rind on basalt lava (eruption of 1907) formed below lichen *Stereocaulon vulcani*. The reddish brown crust contains ferrihydrite, the cream-colored material is lichen thalli, and the black material is fresh basalt. **b:** Bare (uncolonized by lichen) surface of basalt lava (eruption of 1907).

a

b

FIGURE 5-2.
Underside of *Flavoparmelia baltimorensis* thallus (chlorite schist substrate) as seen by scanning electron microscopy. **a:** Broken diatom (composed of silica). **b:** Chrysophytes (composed of silica).

FIGURE 5-3.
Mini-watershed under construction for lichen weathering study at Cone Pond, New Hampshire; subsequently smaller mini-watersheds were used. Rock is mica schist covered mainly by crustose lichens.

gested a methodology for further study (Cochran and Schwartzman 1995; see further discussion in chapter 11). Recently, transmission electron microscopy has been applied to characterize the lichen-rock interface, as well as the microbial influence on weathering (Barker et al. 1994; Barker and Banfield 1996).

In another approach to quantifying lichen weathering (Schwartzman et al. 1997), we have collected data on chemical denudation from mini-watersheds (about 0.05 to 1 m²) located on lichen-covered and bare mica schist from near Hubbard Brook, New Hampshire, the classic site for biogeochemical studies on a forested catchment (Likens and Bormann, 1995). Determination of the elemental runoff fluxes from lichen-covered bedrock could potentially provide quantitative estimates of biotic enhancement of weathering over the abiotic weathering regime. Previous field studies of chemical denudation in alpine areas have implicitly included lichen weathering effects because lichens are ubiquitous on exposed rock surfaces.

Mini-watersheds were created with polyurethane foam boundaries to constrain runoff into collection bottles for two summer field seasons (figure 5-3). Sites were selected to avoid throughfall, with precipitation collectors providing rain samples for each storm event. Bare rock sites were created by splitting local bedrock with incipient fractures, and in one case on a surface exposed by treefall. Because all sites were at elevations of less than 1 km, the preferential capture of elements at lichen sites from a fog-derived deposition flux can be excluded. Filtered water (<0.45 micron) samples were analyzed by inductively coupled plasma methods for magnesium, calcium, sodium, potassium, aluminum, silicon, and iron. Experiments equilibrating polyurethane foam with rain indicate insignificant artifact contributions to measured elemental levels in runoff for all analyzed elements, except perhaps calcium.

For each storm event, the ratio of elemental concentrations in runoff, corrected for precipitation levels, from lichen to bare rock sites (R) gives an apparent ratio in chemical denudation rates for lichen to bare rock weathering at the same climatic conditions. Computed R values (in parentheses) for cumulative fluxes for elements with reliable estimates, with the flux derived from weathering being more than 50% of the total flux, include magnesium (11), silicon (4), aluminum (0.3), and iron (0.1). The enhanced flux from lichen weathering is shown in figure 5-4, where the runoff concentrations of magnesium and silicon, corrected for precipitation levels, are plotted as a function of the reciprocal of precipitation for each rain event. That R values of aluminum and iron are both less than 1 is consistent with progressive accumulation of iron and aluminum in the lichen thallus and its associated products of weathering because these two elements have consistently higher concentrations in analyzed lichen relative to the other elements; the same pattern is found in Jackson and Keller's (1970) lichen weathering study.

Assuming a steady-state lichen biomass and elemental composition, the R values for magnesium and silicon represent minimum estimates of lichen enhancement of chemical weathering over abiotic conditions because these derived values neglect the loss of abraded lichen fragments from the sites during storms, as well as compensating new growth (bare rock sites are not sterile, with likely microbial dissolution taking place and an artifact effect from splitting a large rock boulder to create the sites). Lichen weathering enhancements of one or more orders of magnitude are consistent with previous estimates from studies of weathering rind development under lichens of known ages. Peak day temperatures at the lichen thallus and interface with rock were 10 to 20°C higher than air temperatures, as determined by microthermocouple monitoring. This low albedo-induced warming can be significant in thin soil and bare rock weathering. This effect was cited in a study of alpine weathering (Sommaruga-Wograth et al. 1997).

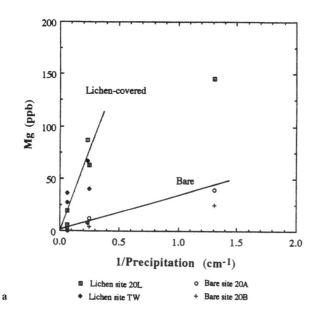

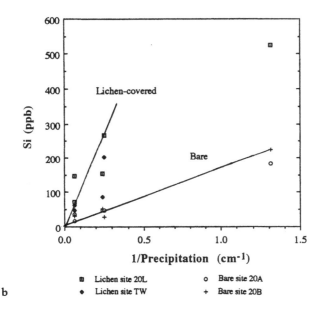

FIGURE 5-4.

a: Concentration of magnesium in parts per billion (ppb) in runoff (corrected for precipitation) from lichen-covered and bare rock mini-watersheds (Cone Pond, New Hampshire) as a function of reciprocal precipitation level for each rain event. **b:** Concentration of silicon in ppb in runoff (corrected for precipitation) from lichen-covered and bare rock mini-watersheds (Cone Pond, New Hampshire) as a function of reciprocal precipitation level for each rain event.

The methodology of using mini-watersheds has real potential to quantify biotic weathering of not only lichens, but also bryophytes and higher plants growing on thin soils, such as in alpine areas.

ALPINE WEATHERING

Several important studies comparing bare rock above the tree line in mountains to forested watersheds below have been undertaken in recent years (Drever and Hurcomb 1986; Mast 1989; Arthur and Fahey 1993; Drever and Zobrist 1992), motivated by the acid deposition problem, as well as the potential insight into biological effects on weathering. However, alpine weathering is not truly abiotic because rock surfaces are microbially colonized by lichens and bryophytes, humus is blown into crevices, and frost wedging, a proposed global biotic effect, is present. Keeping in mind these caveats, these studies have provided very interesting data.

Drever and Hurcomb's (1986) study in the Northern Cascades Mountains looked at weathering of granitic bedrock in the alpine zone. They found negligible weathering of plagioclase (60% of the bicarbonate in runoff was apparently derived from the dissolution of trace amounts of presumably hydrothermal calcite, a phase that is present in many diverse terrains). Arthur and Fahey (1993) compared cationic denudation per unit area of a subalpine forest and alpine zone in north-central Colorado, finding a ratio of forest/alpine equal to 5. 3. Drever and Zobrist's (1992) study in the Swiss Alps, already mentioned, included a similar comparison, measuring cationic and silica denudation rates from catchments on granitic gneiss. They found a ratio of about 25 for cationic denudation of forest to alpine zone. Using an assumed lapse rate and activation energy of 15 kcal/mole, they corrected this ratio for the temperature dependence of weathering, bringing the ratio down to about 8 (Drever 1994). However, as mentioned earlier in this chapter, the temperature correction may not correspond to the actual temperature difference of the weathering interface between the two zones. The actual temperature at the weathering interface during the day in the alpine zone may well be significantly higher than the forests below. One possible inference from these alpine studies is that lichen weathering of rock could produce chemical denudation rates about one order of magnitude lower than forest-covered soil over the same bedrock.

FIGURE 5-5.
Sandbox weathering experiment at Hubbard Brook, New Hampshire. Pine plot is
shown. Courtesy of Ford Cochran.

EXPERIMENTAL ESTIMATES OF BIOTIC ENHANCEMENT OF WEATHERING

At least three recent research programs have looked at microbial and plant
influences on chemical weathering in small-scale experimental set-ups. Two
(Cochran et al. 1996) have utilized the "sandbox" (filled with granitic glacial
sand, underlain by impermeable plastic) at Hubbard Brook Experimental
Forest Station in New Hampshire (Bormann et al. 1987; see Berner 1995a
and Bormann et al. 1998 for further details; see figure 5-5). Over 12 years of
outflow, water samples have been collected from separate plots with red pine
and grasses growing, as well as from a "bare" plot (now colonized by mosses
and sparse lichens) and analyzed. Uptake by biota and dissolution of soil minerals have been studied. Plant growth was estimated and biomass analyzed.
Soil minerals have been studied by scanning electron microscopy, with grains
showing etch pits and secondary mineral formation. Based on sodium and
bicarbonate fluxes, Cochran concluded that at least one order of magnitude
enhancement of dissolution has occurred by pine-induced weathering compared with the bare sand (there were transient effects following rapid spring
uptake by pines). Biomass uptake of sodium, along with the nutrients cal-

cium and magnesium, complicates the interpretation of the results. An enhancement factor of about 10 for calcium of pine tree relative to bare has been estimated (Berner 1997; Berner and Rao 1997). Similar estimates for calcium and magnesium were inferred by taking into account accumulation in secondary minerals in the soils along with bioaccumulation (Bormann et al. 1998). However, it should be pointed out that the sandbox experiment is not really determining the plant (or even moss) biotic enhancement of weathering over the abiotic level because sand is the baseline substrate, with a high surface area/land area, not a more plausible abiotic substrate, a two-dimensional granite platform with little regolith. Nevertheless, these experiments on mini-ecosystems are giving invaluable insight into the biological effects on weathering rates.

The significance of biomass uptake was emphasized in earlier studies (Taylor and Velbel 1991). Taking into account biomass uptake in temperate forested ecosystems has been demonstrated to increase weathering rates of silicates minerals, especially those with nutrient elements (e.g., biotite), by factors of up to 6 or more (Velbel 1995; Taylor and Velbel 1991). This conclusion was based on calculations using stoichiometry of rock-forming silicates and the biomass itself, and observed fluxes in and out of the studied ecosystems. Thus, a material balance of each element could be computed, giving the indicated factor. Only in the case of negative net primary production (degrading ecosystem as a result of fire, infestation etc.) is the biomass factor negative, i.e., a source to the soluble pool.

The third sandbox experiment is currently being conducted (Johansson 1996; Johansson et al. 1997). Preliminary observations on nutrient-enriched crushed granite and basalt kept in a controlled chamber environment for nearly a year indicate a threefold microbial enhancement for basalt and several orders of magnitude for granite, compared with nearly abiotic (nutrient-free) regoliths of the same materials (Johansson measured input and output silica fluxes in these experiments). She concluded that the biotic enhancement resulted from microbial heterotrophic respiration feeding off of autotrophic-derived organic matter. Regoliths with plants are currently being studied in long-term experiments (Johansson et al. 1997).

The possible role of a subsurface microbial community significantly accelerating chemical weathering was highlighted in other research demonstrating very rapid dissolution of granite by bacteria collected in deep environments (Sherriff and Broin 1995). Although one could speculate about microbial involvement in silicate dissolution in soils, particularly in zone C, the transition to fresh bedrock, much more investigation is needed before a definitive case can be made with respect to their significance.

The Laboratory/Field Rate Paradox: Is Biology Irrelevant?

In the chemical weathering community there is little consensus at present on the importance of biota with respect to observed chemical weathering processes. In contrast, soil scientists, geomorphologists, and biologists have generally emphasized the biological role (e.g., Brimhall et al. 1991; Colin et al. 1992; Richter and Markewitz 1995). In their contribution to the excellent survey of the state of the art in chemical weathering studies (White and Brantley 1995), we can thank Blum and Stillings (1995) for making explicit the reservations that some geochemists apparently have with respect to the role of biota:

> Two factors point to biological effects not being a dominant factor: (1) natural weathering rates are slower than predicted laboratory rates determined in abiotic systems, so that any accelerating effects of organisms may be modest, and (2) environments with lower biological activity, such as arctic and alpine environments, do not appear to have dramatically lower weathering rates (p. 335).

Their second point has already been addressed earlier, namely that alpine areas do contain a potential contribution to biotic enhancement of weathering. Their first point, however, deserves serious discussion.

Most investigators agree that dissolution rates of rock-forming silicates measured in the laboratory are two to three orders of magnitude higher than the inferred field rate, using the best estimates of mineral grain size in natural soils (see discussion in White and Brantley 1995; one exception to this consensus is Sverdrup and Warfvinge 1995, who claim they can reconcile the respective rates by systematically estimating all the factors, including the artifact effect, the enhancement of experimental dissolution rates as a result of creating powders). White (1995) pointed out that this discrepancy becomes a factor of 100 to 1000 times larger if the Brunauer, Emmett, and Taylor (BET)-measured rather than the geometric surface area of mineral grains is used in the calculation of dissolution rate (e.g., in units of mole $cm^{-2} sec^{-1}$). The BET method involves the measurement of the volume of gas adsorbed as a monomolecular (or atomic in the case of inert gases) layer on the mineral surface, capturing crevices and internal surfaces not included in a simple geometric surface area (e.g., from mean size of cubic grains). Moreover, the older the soil, the lower the inferred weathering rate from observations of soil minerals. White's explanation is that the reactivity per physical surface area decreases with age, with the most reactive sites (e.g., dislocations) reacting first.

Several explanations have been given for the generally accepted difference between laboratory and field rates:

1. Inhibition of dissolution by solutes (e.g., aluminum) present in soil solutions, but not in comparable levels in laboratory experiments.
2. Protective layer forms over natural mineral grains (Nugent et al. 1998), and the natural mechanism may be very different from laboratory experiments (e.g., see Hochella and Banfield 1995, on clay formation in olivine). Moreover, a progressive decrease in reactive sites occurs as soils age.
3. Soil solutions are generally near saturation (at saturation the dissolution rate is zero because the forward reaction rate equals the back reaction rate) (Drever and Clow 1995).
4. Drying cycles in nature may inhibit dissolution relative to rates in continuously wet laboratory reactors (Brantley and Stillings 1996).
5. Perhaps the most important factor: the estimate of actual reactive mineral surface area in contact with water in the field may be much too high; macropores (channels) in soils divert most of the fluid, thereby bypassing reactive surface, which in turn remains in contact with saturated soil solutions in micropores. Thus, two end members may dominate weathering behavior: under rapid flow, reaction along macropores occurs along with slow transport from micropores, whereas with slow flow conditions the rate is limited by the kinetics of near equilibrium reactions (Drever and Clow 1995).

All of the above factors may well be involved in varying degree in chemical denudation processes on the present Earth, as well as in the past. Does this justify Blum and Stillings' conclusion that "any accelerating effects of organisms may be modest"? Is it still possible that the global biotic enhancement of weathering may be as much as two orders of magnitude? First, laboratory determination of silicate dissolution rates may well be characterized by artifact effects (to be discussed more thoroughly in chapter 6). Sverdrup and Warfvinge (1995) pointed out that many early studies ignored the effect of buffers (organic acids) holding pH constant on the measured laboratory dissolution rate. The soil effect, the increase in potential reactive surface area/ land area, is likely the largest single biotic effect accelerating chemical weathering globally now and in the past. This effect is essentially subsumed in any laboratory experiment that uses a fine powder to determine dissolution rates. This point is not a criticism of laboratory methodology, only of the interpretation of natural weathering. Could other biotic effects, such as direct microbial dissolution of mineral grains in soil, be important? We will return to this question after a full consideration of a parameterization of the physical and chemical influences on weathering rates in the next chapter.

■■ 6 QUANTIFYING THE BIOTIC ENHANCEMENT OF WEATHERING AND ITS IMPLICATIONS

Quantitative estimate of biotic enhancement factor on present Earth surface. Parameterization of the effects of temperature, atmospheric carbon dioxide partial pressure, runoff on weathering rate. My decomposition of the combined effects in chemical denudation. Would the present abiotic Earth be habitable? The biotic factor in GEOCARB. Progress report: what is the cumulative biotic enhancement of weathering?

The role of land biota in affecting chemical weathering rates and the carbon cycle is now under active investigation. Estimates of the present biotic enhancement of weathering range over at least two orders of magnitude. Laboratory and field studies have provided invaluable clues. Another approach has been constructing a model of the temperature/atmospheric pCO_2 history of the Earth's surface based on the silicate-carbonate biogeochemical cycle and testing it with empirical data, a procedure followed for the Phanerozoic by Berner (1993, 1994). We have attempted an analogous approach for Earth history dating back to about 4 Ga (Schwartzman and Shore 1996). This approach will be discussed at length in chapter 8, but it will begin here in model calculations of the likely conditions on an abiotic Earth. We will first look at the basis for a quantitative estimate of the present biotic enhancement of weathering and a systematic parameterization of the various influences on the natural weathering rate. My decomposition of the combined processes leading to typical observed chemical denudation rates in the field will then be discussed. Finally, the biotic factor in weathering will be rejoined in a discussion of GEOCARB.

Quantifying the Biotic Enhancement Factor on Present Earth Surface

The biotic enhancement of weathering factor (B) is defined in a quantitative way as follows: B is how much faster the silicate weathering carbon sink is under biotic conditions than under abiotic conditions at the same atmospheric carbon dioxide level (pCO_2) and surface temperature (Schwartzman and Volk 1989). Although some biotically related effects may locally or temporarily reduce chemical weathering (e.g., macropores in soils, microbial coatings shielding grains), as already mentioned in chapter 5, the net biotic role is likely to increase global denudation rates. There is a consensus that $B > 1$, if only for biotic elevation of pCO_2 in soils (possibly significant even before vascular plants from microbial respiration; see Keller and Wood 1993). Just how much bigger is still under contention. We have argued that B is probably on the order of 100 or even greater. Estimates to date have mainly come from field and laboratory studies of recent weathering.

An indication of the possible magnitude of B comes from the ratio (about 1000) of the rate of chemical weathering in temperate/tropical soil to that of a computed abiotic two-dimensional bare rock rate at similar temperatures and runoff conditions (Schwartzman and Volk 1989), although several caveats are in order (e.g., the likely extent of soil cover and the contribution from porous volcanics under abiotic conditions). This computed ratio of 1000 deserves closer examination. Chemical denudation rates of common silicate rocks in temperate/tropical soils range from 10^{-3} to 10^{-1} mm yr^{-1} with typical values of around 10^{-2} mm yr^{-1} These rates have been estimated from studies of water chemistry of rivers. Two-dimensional dissolution rates of common rock-forming CaMgFe silicates in these rocks are needed for comparison. At the pH corresponding to present atmospheric pCO_2 in equilibrium with water at 25°C (5.66, 3×10^{-4} bars, respectively) the experimental rates are mostly around 10^{-5} to 10^{-6} mm yr^{-1} (hornblende, augite, bronzite, and olivine; the olivine value is from one of the few experiments made on natural unground crystals, from Grandstaff 1986). Diopside, another common CaMg silicate, has a rate that may be as high as $10^{-3.8}$ mm yr^{-1} but an artifact effect may be relevant here (data from Brantley and Chen 1995b). Factors tending to make the ratio of natural soil weathering to two-dimensional bare rock a minimum include the following:

1. The calculation for the two-dimensional bare rock surface assumes continuous flushing with undersaturated water; in nature a bare rock or coarse regolith would not be in continuous contact with water, thereby reducing the computed dissolution rate by the fraction of time there is water contact.

2. The artifact effect in laboratory experiments that measure dissolution rates; grinding of powders may raise measured dissolution rates one or more orders of magnitude as a result of the creation of deformation and dislocations that preferentially dissolve during the dissolution experiment (Petrovich 1981a, 1981b; Eggleston et al. 1989). In particular, Eggleston et al. (1989) found that dissolution rates of diopside powders were an order of magnitude less than initial rates when determined 8 months after laboratory grinding. Is this decline exponential? That is, should we expect in general that silicate powders have experimental dissolution rates that are two to three orders of magnitude higher than for natural weathering under similar pH and temperature regimes? Some experimental data apparently lead to an answer of "no" to this question; no significant difference between dissolution rates was found for some freshly ground silicates and natural mineral powders obtained from sieving soils (Drever and Clow 1995). More studies on aging powders should clarify whether this artifact effect is important. Perhaps the effect is more important for some silicates than others.

Abiotic rock rates for common silicate mineral compositions on the order of 10^{-5} mm yr^{-1} or less is supported by the following studies of natural weathering:

1. A rate of weathered rind development of less than 3×10^{-5} mm yr^{-1} can be inferred from Jackson and Keller's (1970) study on recent Hawaiian basalt flows (this from an area of moderate to heavy rainfall of 127 to 191 cm yr^{-1}).

2. Rates less than 5×10^{-5} mm yr^{-1} are derived from a study of rind development on andesitic and basaltic clasts from the western United States from soils up to 500,000 years old (Colman and Pierce 1981).

Note that in both cases, some microbial involvement is probable, making the rates probable maximums for abiotic rates.

On the other hand, several factors would make the biotic soil/abiotic two-dimensional bare rock rate less than 1000:

1. Some regolith (broken-up rock) would be likely be present on an abiotic Earth surface, particularly in the early Precambrian, when more active volcanism likely ejected volumes of fine ash and more frequent impacts continually pulverized the land surface. What the average or steady-state regolith cover would be (or its thickness and grain size) is a matter for conjecture (further discussion of the possible implications of an abiotic regolith effect is given in the Appendix). However, fine particles would be expected to be rapidly eroded by runoff and wind erosion, leaving a rocky surface, similar to many desert areas. Regolith would probably not form at all in mountainous regions, precisely those sites where the most rapid chemical denudation now apparently occurs because of a combination of mass movement exposing fresh rock and vegetation-anchoring soils. Furthermore, the process of Ostwald ripening, the preferential dissolution of smaller grains relative to larger because of the greater surface area per volume of the former, also may contribute to a reduction of reactive surface area/land area (Steefel and Van Cappellen 1990).

2. Weathering in permeable sediments and volcanics as well as in joints doubtlessly would occur in an abiotic surface regime. However, the role of microbially induced dissolution in these systems may be quite significant now and in the past. Recent research indicates a significant biotic enhancement of dissolution, particularly for volcanic glass, even in marine conditions (Staudigel et al. 1995; Thorseth et al. 1992, 1995a, 1995b). Furthermore, the carbon source for bicarbonate production in many sediments is likely kerogen; thus, the weathering reaction in this case and eventual deposition of carbonate in the ocean may actually constitute a net source of carbon dioxide to the atmosphere, rather than a sink. In any case, this flux of bicarbonate from aquifers is relatively small compared with that carried by the much greater discharge of rivers to the ocean.

In light of the latter factors, the ratio of 1000 may well be an upper limit to B. The lower limit for B, the present cumulative biotic enhancement of weathering on the Earth surface over the abiotic rate at the same temperature and atmospheric carbon dioxide level, is likely close to about 100 (however, even a factor of 10 is quite an enhancement).

Parameterization of the Effects of Temperature, Atmospheric Carbon Dioxide Partial Pressure, and Runoff on the Weathering Rate

An understanding of the possible effect of biota on the habitability of the Earth's surface is dependent on its qualitative and quantitative involvement in the carbonate-silicate biogeochemical cycle. Two different approaches have been taken with respect to modeling the steady-state atmospheric pCO_2/ surface temperature as an outcome of this cycle. One approach has been to parameterize the abiotic and biotic effects as best as can be estimated giving a BLAG or GEOCARB explicit expression of their separate and combined effects on the rate of chemical weathering on a real (biotic) Earth surface. For example, Gwiarzda and Broecker (1994) constructed such a model of current weathering in soil. They included temperature, soil pCO_2 and organic acidity in their model of silicate weathering in a temperate climate, finding that temperature was the main control on weathering rate. Two of their most questionable assumptions were that dissolution occurs far from equilibrium and that microenvironmental effects such as elevation of organic acid concentration and lowering of pH in the rhizosphere are not present. They conceded that microenvironmental effects might be very significant.

The other approach, which we have developed, is to lump all biotic effects, direct and indirect, into a biotic enhancement of weathering factor, with a separate and entirely abiotic parameterization of the effects of atmospheric pCO_2/surface temperature on the chemical weathering rate (Schwartzman and Volk 1989). The original WHAK model essentially treats the kinetics of weathering as an abiotic process with no explicit treatment of biotic effects.

It is important to keep the assumptions in the two distinct approaches in mind, for considerable confusion may result in comparing inferences from each model. For example, one should not conclude from modest incremental biotic effects on chemical weathering rates (e.g., in soil pCO_2) that the absolute biotic enhancement of weathering is small. An abiotic Earth surface could be much hotter than a biotic one, with roughly similar responses to changes in surface temperature and atmospheric pCO_2 levels (i.e., abiotic and biotic cases with similar sensitivities to changes in these parameters). This possibility is analogous to the response of a thermostat at two widely different temperature settings. Analogous misleading conclusions are made from the comparison of field weathering rates to model rates computed from laboratory experiments (see previous discussion in chapter 5).

The physical variables thought to be relevant to the carbonate-silicate biogeochemical cycle include mean and regional surface temperature, land area (at least as a first approximation), volcanic/metamorphic outgassing rate, atmospheric pCO_2, soil pCO_2, and runoff. We will now discuss the parameterization of these variables in modeling.

Atmospheric pCO_2

In an abiotic weathering regime, the H_2O–CO_2 equilibria and kinetic dependence of CaMg silicates on pH will determine the rate of chemical weathering of these silicates for a given reactive surface area, at specified temperature and flushing rate (runoff). The relevant equilibria are:

$$K_1 : H_2O + CO_2 = H_2CO_3$$
$$K_2 : H_2CO_3 = HCO_3^- + H^+$$

(A second dissociation of HCO_3^- to H^+ and CO_3^{-2} will be neglected because the equilibrium constant is much smaller; see Plummer and Busenberg 1982)

Thus, in a system with continuous reaction with water saturated with carbon dioxide and open to the atmosphere,

$$K_1 = aH_2CO_3/aH_2O \, pCO_2$$
$$K_2 = aHCO_3^- \, aH^+/aH_2CO_3$$

For these conditions,

$$aHCO_3^- = aH^+, \text{ thus, } aH^+ = (K_1K_2 \, pCO_2)^{0.5},$$

Hence, aH^+ varies with $pCO_2^{0.5}$. Note that a is the thermodynamic activity of the chemical species.

The rate of dissolution of CaMg silicates varies with $(aH^+)n$, where $0 \leq n \leq 1$; n has been determined for common rock-forming silicates (for pH <7, corresponding to the conditions to be modeled). For inosilicates, n varies from 0.2 (diopside) to 0.99 (augite) (Brantley and Chen 1995), for neosilicates 1.0 (e.g., olivine; Sverdrup 1990), for tectosilicates, particularly plagioclase less calcic than anorthite 0.5 (anorthite, $n = 1$) (Blum and Stillings 1995). Assuming $n = 0.5$ gives $\alpha = 0.25$ (Berner 1992) in the following expression: dissolution rate varies with $(Pab/Po)^\alpha$.

A more rigorous treatment would require the weighted average of n derived from the relative abundances and dissolution rates of CaMg silicates in the exposed continental crust. Given the experimental range of n for these minerals, α probably falls between 0.25 and 0.5 for an open weathering regime. As Berner (1992) pointed out, the open regime would likely occur in weathering above the water table, with continuous resupply of water saturated with carbon dioxide. A closed weathering regime, occurring below the water table, is postulated to result in $\alpha = 1$ because the rate of dissolution would then depend directly on the carbon dioxide concentration of percolating rain water consumed faster than resupply from soil gas or incoming rain water.

What would be the effective α for an abiotic land surface? For bare, soil-free, relatively impermeable plutonic and metamorphic rocks, most of the chemical weathering could well take place in an open system regime with $0.25 \leq \alpha \leq 0.5$. Weathering of porous volcanics and sediments might entail a range of α depending on where most dissolution takes place, above or below the water table (e.g., apparently, α is close to 1 for spring waters draining basalts in Iceland; see Gislason and Eugster 1987). Other possible influences on dissolution include buildup of bicarbonate in soils, corresponding to $\alpha \leq 0.5$, but at higher pH, near 7, hence in the minimum region of dissolution. For most rock-forming silicates the dissolution rates increase as pH drops below about 5 and increases above about 8; thus, a dissolution minimum exists around neutrality (White and Brantley 1995). In alkaline solutions, dissolution occurs by OH^- complexing with surface cations.

Berner (1992) pointed out that an abiotic land surface would likely be much drier, with windblown alkaline dust neutralizing rain acidity, again tending to minimize dissolution, requiring still higher atmospheric pCO_2 for a stronger greenhouse effect, hence higher surface temperatures, for attaining a weathering sink equal to the volcanic/metamorphic source. In addition, ". . . to avoid unreasonably high CO_2 levels, and excessive global

warming, one is forced to choose the closed-system, linear-feedback option
. . . There are additional complications in a barren world. One is that the
absence of plants would vastly change evapotranspiration. . . . Because of a
lack of evaporation there would be lower cloudiness and much less rainfall
over the continents . . . leading to large-scale desertification and thus, lower
rates of weathering" (Berner 1992, p. 3227). As we will see in subsequent
chapters, these arguments are consistent with just such a high atmospheric
pCO_2 (and temperature) regime on Archean/early Proterozoic continental
surfaces, with limited microbial colonization. An open system weathering
regime is not an implausible approximation for this scenario.

Temperature and Runoff

The direct dependence of chemical weathering of CaMg silicates is assumed
as follows:

$$R/R_o = e^{\beta \Delta T}, \text{ after Walker et al. 1981,}$$

where R is rate at T, R_o at T_o, $\Delta T = T - T_o$, T in °K. This is the standard
Arrhenius expression applicable to chemical reactions.

The exponential term is a simplified version of the corresponding term in
the Arrhenius relationship:

$$e^{E/R[(1/T_o)-(1/T)]},$$

where E is the reaction activation energy and R is the gas constant.

Note that

$$[(1/T_o) - (1/T)] = (T - T_o)/T_o T, \text{ with } \beta = (E/RT_o T).$$

Assume $\beta = 0.056$ (same as Walker et al. 1981), which corresponds to $E =$
50 kJ/mole (11.9 kcal/mole) and $TT_o = (285) (385)°K^2$, $R = 0.008314$ kJ
mole^{-1} °K^{-1}. Others (e.g., Marshall et al. 1988) have assumed lower values
of β (0.046). Most experiments on mineral dissolution made near neutral
pH give $E \geq 50$ kJ/mole (Brantley and Chen 1995). As pH drops below 7, E
apparently increases (Casey and Sposito 1992). Berner (1994) adopts Brady's

(1991) preferred value of 15 kcal/mole = 63 kJ/mole for CaMg silicates in GEOCARB II.

There have been several attempts to derive a temperature dependence of silicate weathering from field data. Velbel (1993) derived an apparent activation energy of 77.0 kJ/mole from the temperature dependence of natural plagioclase weathering in two catchments at different elevations, which were apparently similar in other respects. However, the small temperature difference of 1.1°C between catchments, derived from an assumed lapse rate, may be too small to put much faith in the derived activation energy. In a much larger sample of watersheds ($n = 68$) underlain by granitoids, White and Blum (1995) derived apparent activation energies of 59.4 and 62.5 kJ/mole from silica and sodium flux variations with temperature, respectively. One question remains unresolved: are these activation energies applicable to abiotic weathering of silicates, or are field-derived activation energies in some way biologically affected through some intricate feedbacks between hydrology, temperature, biotic productivity, organic acid production, and soil pCO_2, or even microbial colonization of primary minerals in soils? The same study also derived a relationship between weathering fluxes of silica and sodium as a couple product of both precipitation and temperature:

$$Q = (a \star P)e^{-E/R[(1/T)-(1/T_0)]}$$

where the $(a \star P)$ term assumes a linear correlation between precipitation (P) and fluxes. Runoff is roughly correlative with precipitation in the watersheds studied, with the difference between precipitation and runoff being evapotranspiration. The silica flux correlates similarly with runoff and precipitation.

Although vegetation is not explicitly examined as a weathering factor, the researchers suggested that "any effect of vegetation may be incorporated into the climatic trends."

The use of global mean surface temperatures (averaged air temperatures at the approximate height of 2 m above the surface) in modeling is a first approximation that may need correction in application to models of global climatic regulation. Specifically, ground and soil temperatures may depart significantly from air temperatures. Temperature fluctuations may produce significant nonlinear effects on weathering; that is, short periods of relatively high temperature will have a greater potential effect on fluxes than long periods of low temperature; weathering deeper in the soil profile, with smaller temperature fluctuations, may be approximated more closely by the mean annual temperature (Velbel 1990; Lasaga et al. 1994).

Even more dramatic effects may occur on bare or lichen-colonized rock, on soil surfaces with low albedos, at low air temperatures at high elevations. Here, peak day ground temperatures are commonly higher by 20°C or more than air temperatures (Geiger et al. 1995). This phenomenon may be very relevant to temperature corrections on inferred weathering parameters in alpine terrains. For example, as previously mentioned, Drever and Zobrist's (1992) temperature correction for the alpine zone weathering may be the opposite of what is appropriate, resulting in even a higher temperature-corrected ratio between forested to alpine flux than computed from solute concentration ratios.

Finally, latitudinal placement of continents may have strong effects on weathering rates because land mass in the tropics weathers more intensely than at high latitudes (Worsley and Kidder 1991), an effect not captured by using global mean temperatures in modeling (White and Blum 1995). However, if the mean global surface temperature (Tm) was significantly higher more than 1.5 billion years ago, as will be argued in chapter 8, then latitudinal differences in temperature likely would have been much smaller or even nonexistent, making the first approximation use of Tm for that time much more appropriate (the possible complication of normalization to the present still remains; see later discussion on this point).

Runoff dependence on Tm is assumed to vary as $e^{(\gamma \Delta T)}$, $\gamma = 0.017$ (Walker et al. 1981, after Manabe and Stouffer 1980 and Manabe and Wetherald 1980, whose climate models looked at regional dependence of runoff on temperature and simulated global dependence as a function of global mean temperature; Berner 1994 used a similar parameterization). Stronger dependences have been proposed (e.g., Sellers 1965), but the limit $e^{(\gamma \Delta T)} \leq 2$ is used because of energy constraints limiting maximum global runoff, The exponential increase in the rate of evaporation and hence precipitation with increasing temperature is ultimately limited by the available solar flux to heat the ocean surface. Because about one half of the solar flux absorbed by the oceans is now supplying the latent heat of vaporization, the limit on evaporation/precipitation is roughly twice the present (Pollack et al. 1987). It is further assumed that the global weathering rate is proportional to the runoff term. The choices of β and γ are somewhat arbitrary, but what counts is their effect on the computed parameters, as we will shortly see.

My Decomposition of Combined Effects in Chemical Denudation

An outline of a hypothetical decomposition of effects that generates the observed field rate is shown in table 6-1.

The estimates of a typical field chemical denudation rate and a computed
two-dimensional rate for common rock-forming CaMg silicates have been
previously discussed. The estimated magnitudes of each effect in table 6-1
are, at best, informed "guesstimates." The hydrologic factors are presumed
to be the macropore effect of most water flushing through a soil bypassing
reactive mineral grains, along with the apparently common condition that
most of the soil water is usually close to saturation. The latter effect would
reduce dissolution rates because reactive mineral surface in contact with saturated solution will not dissolve, being at chemical equilibrium. The combined hydrologic effects are presumed to reduce dissolution rates by a factor
of 10^{-1} to 10^{-3}. The inferred effect of pCO_2 soil elevation is perhaps the
most robust because pCO_2 soil levels have been extensively measured. A
factor of 3 to 6 is computed, using the previously derived expression, in this
case applied to a realistic soil on present Earth, i.e., R is proportional to
(pCO_2 soil/pCO_2 atmosphere)$^{\alpha}$. The soil effect assumes a plausible combination of a range of grain size of CaMg silicate minerals and thickness in a
soil; for the example previously given, 1 mm grain size, 1 m thick, gives an
elevation in surface area over a two-dimensional surface of 6000 times. The
microenvironmental effects of enhanced dissolution from mineral contact
with roots, with organic acid levels and lower pH than bulk soil solution, are
probably the biggest guess because little research has focused on quantifying
these effects and their impact on the observed chemical denudation rate. I
would guess a factor of 10 to 10^2.

The multiplied result of all these factors ranges from 3×10^2 to 6×10^3
compared with the ratio of field chemical denudation rate to computed two-
dimensional rate ranging from 10^2 to 10^4. Therefore, these two ratios agree.
I will not make too much of this agreement given the uncertainties involved.
Of all the above effects, the magnitude of the hydrologic and microenvironmental effects are the most uncertain. However, given these uncertainties,
there is little justification to claim, as Drever (1994) did, that "there is not
an orders-of-magnitude acceleration attributable to microenvironments." In
conclusion, I claim that a plausible decomposition of combined effects occurring in soils can be made that is consistent with relatively high biotic

TABLE 6-1.
Decomposing Combined Effects in Chemical Denudation

Effect	Estimated Magnitude
Hydrology: macropore (bulk of soil water near saturation)	10^{-1}-10^{-3}
Microenvironment: organic acids, chelation, direct microbial dissolution of silicate grains	10-10^2
pCO_2 soil elevation (α = 0.25-0.4, 100 PAL in soil)	3-6
"Soil effect" (increase in potentially reactive surface area/land area)	10^2-10^4

Range of combined effects: 3×10^2 to 6×10^3, compared with $(R/R_{2D}) = 10^2$–10^4
The proposed rate of chemical denudation (R) is assumed to be 10^{-2} mm yr^{-1} (tropical/saprolitic denudation of granitic rock).
The two-dimensional geometric rate for CaMg silicates (R_{2D}) at 1 PAL CO_2, 25°C, pH = 5.5): 10^{-4}–10^{-6} mm yr^{-1} (assumes roughness factor of 10, i.e., the ratio of BET to geometric surface area; see White 1995)

enhancement of weathering. This should be taken as a challenge to the weathering community to include this possibility in their research program. Only detailed studies of specific field sites will clarify the relative importance of each postulated effect and their combined influence on observed chemical denudation rates in soils and watersheds.

A Present Abiotic Earth: Would It Be Habitable?

Let $B = W_o/W_{abiotic}$ where W_o is the present global biotically enhanced chemical weathering rate and $W_{abiotic}$ is the hypothetical abiotic rate at the same atmospheric carbon dioxide level, surface temperature, V, and A as present. Then,

$$(1) \quad (V/V_o)W_o = W_{abiotic}(P_{ab}/P_o)^\alpha \, e^{(\beta \Delta T)} e^{(\gamma \Delta T)}(A/A_o)$$

where A_o, A and V_o, V are the present and arbitrary different continental land areas and volcanic/metamorphic outgassing rates of carbon dioxide, respectively. P_o is the present atmospheric partial pressure of CO_2, P_{ab} the partial pressure of CO_2 required under abiotic conditions for steady state, ΔT is the temperature elevation required for an abiotic condition over the present global mean, T_o, taken as 288°K ($\Delta T = T_{ab} - T_o$; T_{ab} is the abiotic surface temperature). As previously discussed, the factors α, β, and γ express the dependence of silicate weathering rate on pCO_2, temperature, and runoff, respectively; assume $\beta = 0.056$ and $\gamma = 0.017$, with $e^{(\gamma \Delta T)} \leq 2$.

Looking at the meaning of equation (1), increasing A will increase the abiotic weathering rate (the whole expression on the right hand side of the equation) with no necessary change in other parameters, whereas increasing V will require an increase in the other physical parameters for abiotic weathering to balance this higher outgassing rate.

Rearranging and simplifying,

$$(2) \quad B/\{(A/A_o)(V_o/V)\} = (P_{ab}/P_o)^\alpha e^{(\beta \Delta T)} e^{(\gamma \Delta T)}.$$

Note that for $B = 1$ (i.e., no biotic enhancement of weathering), $W_o = W_{abiotic}$, and all abiotic parameters equal their biotic counterparts. This formulation adds the biotic enhancement term B to the carbonate-silicate model developed by Walker et al. (1981) and assumes a steady state between the silicate weathering carbon sink and volcanic/metamorphic source with respect to the atmosphere (Schwartzman and Volk 1989). It is important to note that all biotic influences on chemical weathering (soil pCO_2 elevation, soil stabilization, microbially induced dissolution, etc.) are subsumed in the B term because it is defined as the ratio of the global biotically enhanced rate to the hypothetical abiotic rate. As such, the present value of B is the cumulative result of biospheric evolution since the origin of life.

We now consider the implications of a present biotic enhancement of weathering of up to 1000 times over the abiotic rate using a standard technique for quantitatively estimating effects on the carbonate-silicate geochemical cycle. We balance today's metamorphic and magmatic outgassing of CO_2 (V_o) with the CO_2 sink from global silicate weathering, following a similar parameterization as Walker et al.'s 1981 paper.

This parameterization including V/V_o, A/A_o is not needed in considering the present Earth but will be important in modeling the past (to be addressed in chapter 8).

The biotic enhancement, B, is varied between 1 and 1000. We present results using two different greenhouse calculations that assume different dependencies of surface temperature on atmospheric pCO_2 level. The results in figure 6-1 show that if the biotic enhancement ratio $B = 10$, then a

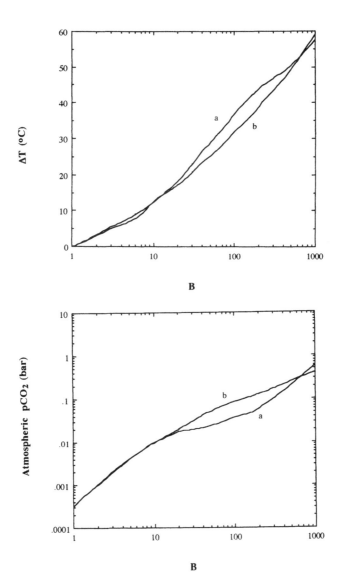

FIGURE 6-1.
The temperature elevation (**top**) over the present mean global temperature and corresponding atmospheric pCO_2 level (**bottom**) on a present-day abiotic Earth surface as a function of the enhancement of the present biotic weathering rate relative to an abiotic rate (B) under the same atmospheric pCO_2 and surface temperature conditions. Computed results for two different greenhouse functions (a and b) are shown.

TABLE 6-2.
Computed Present Biotic Enhancement of Weathering

T_{ss} (°C)	B ($E = 50$ kJ/mole)		B ($E = 63$ kJ/mole)[a]	
Assumed	$\alpha = 0.3$	0.4	$\alpha = 0.3$	0.4
60	164	279	342	580
55	100	158	194	307

These results used the updated Caldeira and Kasting greenhouse function (see Chapter 8 for explicit expression).

[a]Computed from factor $= e^{(E/R)[(1/T_o) - (1/T)]}$ instead of simplified parameterization $e^{B(\Delta T)}$

present abiotic Earth surface would be 25 to 30°C; for $B = 100$, the corresponding values are 35 to 55°C; and for $B = 1000$, the probable upper limit, the values are 50 to 75°C using the greenhouse functions of Walker et al. (1981) and Kasting and Ackerman (1986). In other words, to get the present, biotically enhanced, global weathering rate balancing the volcanic/metamorphic source requires higher temperatures and atmospheric pCO_2 levels on an abiotic Earth, doing the "work" of biology. Note that a surface temperature of 50°C is the upper temperature limit for the growth of Metazoa and most other eucaryotes except for thermophilic fungi and a few hardy protists with an upper temperature limit of 60°C, the upper limit of complex life. However, even for the probable upper limit of $B = 1000$, the Earth would be habitable for thermophilic procaryotes. These results were originally published in Schwartzman and Volk (1989). These conclusions are basically substantiated using a more precise formulation of the direct temperature dependence of chemical weathering and a more plausible value of its activation energy for abiotic weathering of CaMg silicates (i.e., $E = 63$ kJ/mole instead of 50 kJ/mole), except that somewhat higher values of B are required for given temperatures on a present abiotic Earth (table 6-2).

A pioneering quantitative approach to estimating temperatures on a present abiotic Earth was made by Lovelock and Watson (1982), assuming the diffusion of atmospheric carbon dioxide into soil. They calculated a temperature increase of about 20°C corresponding to atmospheric $pCO_2 = 0.01$ bar, or roughly the present biotic soil level, for the weathering sink to balance volcanic emissions. Although they recognized that the pCO_2 and temperature would be somewhat higher because of a concentration gradient in the soil, the model did not include the temperature dependence of weathering, which would lower the required pCO_2 and surface temperature. In any case, simply substracting out the biotic effect of soil pCO_2 elevation only goes part of the way to simulating abiotic conditions (e.g., the likely absence of soil stabilization and other factors that constitute the biotic enhancement of weathering need to be taken into account).

We return to the question of uninhabitability for complex life in chapter 10, where similar calculations will be presented with respect to understanding the width of the habitable zone on Earth-like planets.

The Biotic Factor in GEOCARB

As already pointed out, GEOCARB and models like it explicitly incorporate a parameterization of the biotic influence on weathering rate. Berner (1991, 1994) postulates an evolutionary increase in the biotic impact on weathering, starting with primitive microorganisms, culminating in the vascular land plants dominated by angiosperms. Before the rise of vascular land plants, he assumes that $\alpha = 0.5$ in the term $(pCO_2t/pCO_2now)^\alpha$, with α intermediate between open and closed weathering, with pCO_2 being the atmospheric levels. Between 350 and 300 million years before present (BP), this factor and a vascular plant factor are linearly mixed in their impact on silicate weathering, with the latter corresponding to a carbon dioxide fertilization factor, $[2 \, Pr/(1 + Pr)]^{0.4}$ (where Pr is the ratio of pCO_2 at time t to now), a formulation including the presumed effect of faster rock decomposition from both soil microbial weathering and a biotic sink effect. This formulation was taken from Volk (1987). It assumes that plant growth is enhanced by rising carbon dioxide levels in the atmosphere, a concept with some support from experiments and modeling. The exponent 0.4 was set somewhat arbitrarily according to Berner (1991) to take into account that not all plants respond to carbon dioxide fertilization (see Berner 1991 for further details). In addition, another biotic factor, $fE(t)$, is used in the modeling: $fE(t) = 1$, 80 million years BP to now, $fE(t) = 0.15$ prior to rise of plants (starting 300 million years BP). Between 300 and 130 million years BP, $fE(t) = 0.75$ (pteridophyte and gymnosperm dominated).

The prevascular factor of $fE(t) = 0.15$ is supported by its similarity to the ratio (0.13) derived from a study of contemporary weathering in the Swiss

Alps, with temperature-corrected weathering rates in forested and alpine terrains being compared (Drever and Zobrist 1992), but see the previous discussion of this correction. This agreement is probably coincidental for several
reasons. First, the ratio of biotic factors of the prevascular to angiosperm
used in GEOCARB II is not 0.15 but rather 0.47 if the ratio of the full
parameterization is computed:

$$\text{prevascular/angiosperm} = [P_r^{0.50} \times 0.15]/[2P_r/(1 + P_r)]^{0.4} = 0.47,$$

here prevascular $pCO_2 = 16$, the preferred paleolevel, just prior to the emergence of vascular plants some 450 million years BP. Second, one should not
expect agreement considering the likely differences in biota and other factors
between early Paleozoic weathering (probably dominated by bryophytes)
and contemporary alpine weathering promoted by lichens, bryophytes, and
so forth; humus blown into crevices; and frost wedging, the latter a possible
global biotic effect.

Nevertheless, Berner's prevascular factor of $fE(t) = 0.15$, is consistent
with a ratio of biotic enhancement of weathering factors of about one order
of magnitude for the present vascular plant-dominated ecosystem to the
primitive Precambrian land biota (the biotic enhancement of weathering
is defined for the same temperature and atmospheric pCO_2 conditions,
whether comparing biotic and abiotic or two different biotic regimes). Further discussion of this issue is found in the section on estimates of the current
biotic enhancement of weathering from field studies (chapter 5).

Progress Report: What Is the Cumulative Biotic Enhancement of Weathering?

If the biotic enhancement of weathering factor of vascular plants, lichens,
and bryophytes is one order of magnitude greater than primitive land biota
of the Precambrian, an estimate consistent with field, experimental studies,
and GEOCARB modeling, then only another factor of 10 is required of the
latter over an abiotic regime to give $B \approx 100$, the order of magnitude we
infer from inverting the Archean/early Proterozoic record from modeling
the carbonate-silicate cycle, assuming a high surface temperature (50–70°C)
scenario and a carbon dioxide greenhouse (to be discussed in chapter 7). An

TABLE 6-3.
Computed Effective Biotic Enhancement Factors

			computed B_e values			
R	E	(x:	0	0.1	0.5	1)
1000	1000		1000	9.91	2	1
	100		1000	91.7	19.8	10
	10		1000	526	182	100
100	1000		100	1.0	0.2	0.1
	100		100	9.2	2.0	1
	10		100	53	18	10

The model calculations illustrate that, except for a low assumed effect on weathering intensity from abiotic regolith (E), the computed effective biotic enhancement factor (Be) declines markedly with modest increases in abiotic regolith coverage (x). We have argued here that both relatively small abiotic regolith coverage and weathering intensity are likely as a result of water and wind erosion on bare abiotic land surfaces, implying that relatively high effective biotic enhancement factors should be used in modeling. Perhaps geomorphologic data and theory will provide more insight on the likely regolith weathering effect on abiotic land surfaces.

additional order of magnitude for the net enhancement on land colonized by primitive biota over abiotic appears to be plausible given the range of potential effects on each scale.

Appendix: Abiotic Regolith Effect

The possible implications of the impact of regolith cover on an abiotic land surface to weathering rates is discussed below in consideration of a simple model.

Let E = the weathering enhancement factor from abiotic regolith (i.e., enhancement of weathering intensity, flux/land area); x = fraction of land covered by regolith; R = ratio of biotic weathering intensity (soil of course included here)/abiotic intensity on two-dimensional land surface; B_e = effective biotic enhancement factor = (biotic weathering intensity)/abiotic (with regolith).

Let a_o = abiotic weathering intensity with no regolith and a = abiotic weathering intensity with some regolith coverage. Then,

$$a = a_o (1 - x) + E(x) \text{ and } R/B_e = a/a_o$$

Thus,

$$B_e = (a_o R)/[x(E - a_o) + a_o]$$

The relevant functionality is between B_e and x. Let a_o = 1. The results of the model calculations are shown in table 6-3.

A reinterpretation forced by the challenge to the reality of ancient glaciations (Oberbeck, Rampino): tillites as impactites. A much warmer Earth for most of geologic time than heretofore assumed. A proposed first-order temperature curve for the Earth's surface: a summary of evidence from the geologic, paleontologic, and geochemical record (fossil record/molecular phylogeny, paleotemperatures from oxygen isotopes in sedimentary rocks, a temperature limit from gypsum precipitation? glaciations?).

According to paleoclimatologists, the long-term surface temperature history of the Earth is constrained by the geologic and fossil record and plausible inferences about the history of the Earth-sun system (see chapter 3). Inferences about surface temperatures and climate come from a wide variety of observations. Helpful reviews are provided by Crowley and North (1991) and Frakes et al. (1992). The most reliable interpretation comes from the most recent record, especially the past 20,000 years. For example, the analysis of air bubbles trapped in Antarctic ice of known age provides a detailed account of atmospheric carbon dioxide fluctuations with oxygen and hydrogen isotopic composition of the enclosing ice providing a paleotemperature record of extraordinary accuracy for the past several hundred thousand years (a good summary of this evidence was provided by Skinner and Porter 1995).

Evidence for temperature/climate history comes primarily from the isotopic, mineralogic, and fossil records preserved in sedimentary rocks. The oxygen isotope record is considered a fairly reliable indicator for the past 100 million years. Before that time, uncertainties in interpretation arise because of possible variation in the oxygen isotopic composition of seawater and burial effects that may have altered the primary isotopic signature derived from the equilibration of precipitating mineral phases (in chert or carbonate sediment) with seawater. These complications, however, do not necessarily mean that valid data on ancient climate cannot be obtained, as we shall soon see.

Turning to other aspects of the sedimentary record, the mineralogy and lithology, as well as the fossils themselves, can provide valuable insight into ancient climates. Evaporite (e.g., rock salt) deposits are generally taken to indicate warm conditions by analogy with their present environment of deposition (e.g., salt flats or sabkas of the Persian Gulf). Tillites, or sediments resembling Pleistocene and recent glacial deposits, are taken to indicate cold conditions. Fossil soils (paleosols) can give insight into ancient climate from both their mineralogic and isotopic compositions. For example, clay mineralogy can indicate deep tropical weathering. The carbon isotopic composition of carbonates and iron oxides with combined carbonate (goethite) can be used to infer ancient atmospheric carbon dioxide levels, as far back as the early Paleozoic (see chapter 2). Finally, mainly by analogy with contemporary flora and fauna, but also clues provided by the host rock, the fossil record provides valuable data for reconstructing ancient climates. For example, growth rings in petrified wood from the Cretaceous can provide evidence of seasonality of climate (Frakes et al. 1992).

A synthesis of the history of the mean global surface temperature for the Phanerozoic is shown in figure 7-1 (after Frakes et al. 1992). The estimated limits of the shifts from warm to cold conditions are 5 to 10°C from the present global mean of 15°C. Now turning to the surface temperature history going back to the early Archean, we first consider the conventional history with ambient temperatures not markedly different from the Phanerozoic. This conventional history, particularly for the Precambrian, is based on its presumed glacial record and inferred precipitation of primary gypsum, a calcium sulfate mineral found in evaporite deposits. However, the inertia of uniformitarianist thinking should not be underestimated as a factor in cementing the conventional consensus.

In particular, the apparent glacial epochs in the Precambrian, the earliest generally accepted record being the Huronian (dated at about 2.3 Ga) has been taken to place an upper limit of about 20°C on the global mean temperature at those times (Crowley 1983; Kasting and Toon 1989). One other constraint has been commonly cited, from the earliest Archean sedimentary

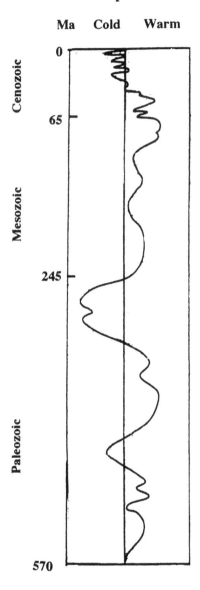

FIGURE 7-1.
Estimated mean global temperature curve for the Phanerozoic. (After Frakes et al. 1992.)

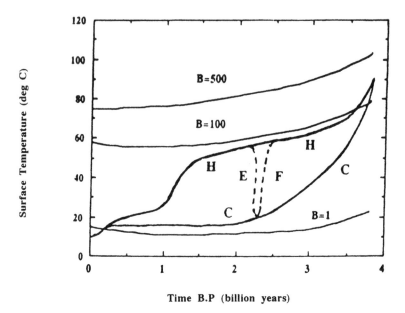

Time B.P (billion years)

FIGURE 7-2.

Earth surface temperature versus time BP (billion years). Two alternative scenarios are noted (*H* and *C*, with possible excursion *E–F*). Temperature trajectories for $B = 500$, 100, and 1 are also indicated (for preferred model b); see discussion in Chapter 8.

record, the inference of primary precipitation of gypsum from either fresh water ($\leq 58°C$) or seawater ($\leq 18°C$) (Walker 1982). The first-order surface temperature curve of the Earth, leaving out fluctuations on the order of 5 to 10°C, is shown as curve C in figure 7-2. In our earlier papers (Schwartzman and Volk 1991a, 1992), we uncritically accepted this consensus (Crowley and North 1991), in spite of already published contrary indications that surface temperature could have been significantly higher, derived from data on apparent paleotemperatures from Precambrian cherts (Knauth and Epstein 1976; Karhu and Epstein 1986).

Kasting (1992) put it as follows in his contribution to the monumental survey of the Proterozoic, referring to the Karhu and Epstein study:

Unfortunately the temperatures inferred from the isotopic data exhibit no correlation with the climate history suggested by the glacial record. It therefore seems likely that the isotopic data reflect temperatures during diagenesis rather than during deposition (p. 165).

What triggered my rethinking of the conventional scenario (curve C) was becoming aware of a radically new interpretation of the geological record of ancient glaciations. In November 1992, Michael Rampino told me of a hypothesis he had been working on for several years, namely that at least some of the assumed record of ancient glaciations were impact deposits. (It turned out that Verne Oberbeck had published this idea the previous spring, a fact unknown to Michael at that time.) Whether this reinterpretation is correct, it led me to a reinterrogation of the assumed constraints on temperature, particularly for the first two thirds of Earth history (Archean, early Proterozoic). This rethinking led as well to the revival of a hypothesis first articulated by Hoyle (1972; mentioned by Knauth and Epstein 1976), namely that a warm early Earth held back the emergence of low-temperature life (this hypothesis is discussed at length in chapter 9). This reinterrogation resulted in a proposal that the temperature trajectory (curve H in figure 7-2) for the Earth's surface was much warmer for most of the Precambrian than the conventional curve. We will now systematically examine the case for a very warm (50–70°C) Archean/early Proterozoic (3.8–1.5 Ga).

Fossil Record and Molecular Phylogeny

Microbial fossils of this age are consistent with thermophily, based on very close morphological similarities with living procaryotes (Schopf 1992) which have thermophilic varieties (e.g., Chroococcales, Chloroflexus, Oscillatoriales, Nostocales) (figure 7-3). The remarkable correspondence of the sequence of appearance of life forms (see table 8-1) in the Precambrian to their upper temperature limits led Hoyle (1972) to suggest that hot surface conditions on Earth may have held up the emergence of complex life. The thermophilic character of deeply rooted Bacteria and Archea (figure 7-4) is consistent with very warm conditions on the Archean surface (Woese 1987; Spooner 1992), but does not require them because early thermophilic forms may have been restricted to similar local environments as now (i.e., hot springs, thermal vents). This universal phylogenetic tree depicted in figure 7-4 is based on the detailed comparison of ribosomal RNA sequences pioneered by Woese (1987). Those lineages closest to the root are hyperthermophilic (i.e., grow above 85°C). Present-day thermophilic communities could be living models for a surface biota inhabiting both oceanic and

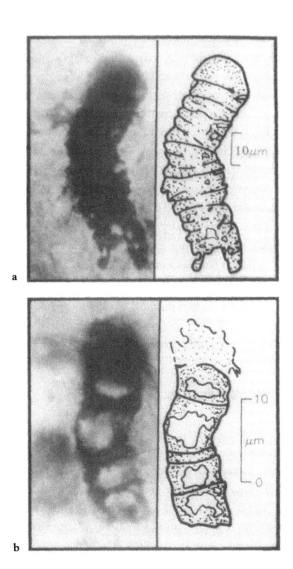

FIGURE 7-3.
A selection of microfossils (with interpretative drawings) shown in thin sections of the early Archean (3.5 Ga) Apex chert of Western Australia. Magnification is shown by scale. **a:** *Archaeoscillatoriopsis maxima.* **b:** *Primaevifilum laticellulosum.* (Courtesy of Bill Schopf.)

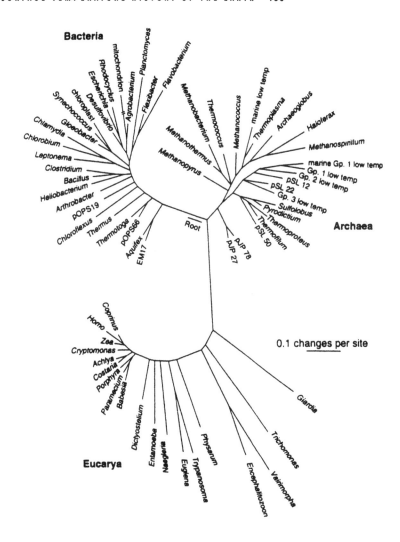

FIGURE 7-4.
Universal phylogenetic tree based on rRNA sequences. (After Pace 1997.) The scale bar corresponds to 0.1 changes per nucleotide.

terrestrial environments in the Archean/early Proterozoic if thermophily was required at that time. As Knauth (1992) put it, the Earth's surface was an "open hydrothermal system in the Archean."

There is now some evidence that a thermophilic eucaryotic algae, Cyanidoschyzon, now living in CO_2-charged water up to 57°C (Seckbach 1994), may be a living model of a primitive photosynthesizing eucaryote emerging some 2 to 2.5 billion years ago. Seckbach (1994) originally suggested that Cyanidoschyzon was a model for primitive mitochondrial eucaryotes, but recent phylogenetic evidence supports it being rooted on the tree after ancentral protozoa (Seckbach 1997).

Paleotemperatures from the Oxygen Isotopic Record of Chert and Carbonates

Paleotemperatures have been derived from the oxygen isotope record of pristine Precambrian cherts (Knauth and Epstein 1976; Karhu and Epstein 1986; Knauth and Lowe 1978; Knauth 1992) and carbonates (Burdett et al. 1990; Winter and Knauth 1992). These inferred paleotemperatures are quite high in the Archean/early Proterozoic, lower in the late Proterozoic and Phanerozoic (figure 7-5). Given the estimated uncertainties in the inferred climatic temperatures of 5 to 10°C (Knauth, personal communication), this temperature history is essentially identical to scenario H in figure 7-1. The following assumptions underlie the computation of climatic paleotemperatures.

The oxygen isotopic composition of seawater at the time of formation was the same as today (corrected by melting the icecaps). This assumption of constancy received support from Holmden and Muehlenbachs (1993), based on the results of their study of a 2 billion year old ophiolite (ancient seawater altered oceanic crust), from which they inferred the isotopic composition of contemporary seawater since the primary signature of the oceanic crust was known. Isotopic studies of Archean greenstones and hydrothermal deposits also yielded results consistent with near constancy back to 3.5 Ga (Holmden and Muehlenbachs 1993). This constancy is explained by the continuous reaction of seawater with oceanic basalts at 200 to 300°C (Muehlenbachs and Clayton 1976); the isotopic exchange constrains the steady-state oxygen isotopic composition of seawater. Similar conclusions are derived from measurements of oxygen isotopes in hydrothermal fluids from mid-oceanic ridges (Jean-Baptiste et al. 1997). The latter authors support the interpretation of high Precambrian ocean temperatures from the chert and carbonate record.

However, paleotemperatures obtained from the fractionation of oxygen isotopes between coexisting chert and inclusions of phosphate mineral are

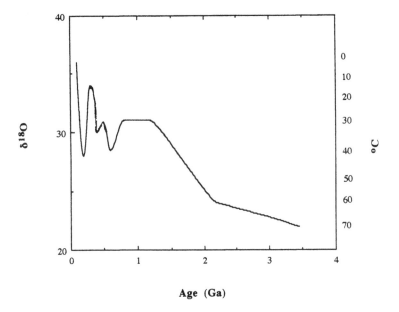

FIGURE 7-5.

The highest measured $\delta^{18}O$ of cherts as a function of age. Cherts known to be altered by metamorphism have been omitted from the data. These values at any age best approximate $\delta^{18}O$ of chert in isotopic equilibrium with sea water. Assuming that $\delta^{18}O$ of seawater was approximately constant at -1 per mil, surface temperature variations over geologic time are derived from this curve (shown on right-hand scale). (After Knauth 1992).

apparently independent of the seawater oxygen isotopic composition (Karhu and Epstein 1986). The climatic temperature history derived from this data set agree with those from chert alone, with the assumptions described here.

Temperatures derived from the "heaviest" (i.e., highest $^{18}O/^{16}O$ ratio) of the Precambrian cherts (or carbonates) are set in early diagenesis near the seawater/sediment interface and as such approximate seawater and hence climatic temperature (see Knauth 1992). The high temperatures are commonly interpreted as representing conditions during deep burial or metamorphism (see previously cited quote from Kasting 1992; Perry 1990, with reply by Epstein and Karhu, 1990). The inference of actual climatic temperature is supported by a number of arguments. First, the heaviest oxygen in these Precambrian samples occurs in unrecrystallized cherts and carbonates (which also preserve pristine textures and facies trends in both carbon and oxygen isotopes). The only way to reset the oxygen in the cherts is by an even higher temperature fluid percolating through the rock, with a high water/rock ratio. This process would likely recrystallize the mineral grains in the rock (see Jones and Knauth 1979). Second, Precambrian nodular cherts often show sedimentary structures consistent with crystallization near the water/sediment interface (Knauth and Clemens 1995).

Knauth and Clemens (1995) have recently reported a large new data set that includes unmetamorphosed cherts with microfossils. Their results are consistent with the inferred Precambrian temperature history given by Knauth (1992), shown in figure 7-5.

There is one report in the literature of a chert claimed to be of Archean age, with primary sedimentary textures, giving a range of $^{18}O/^{16}O$ ratios that would correspond to temperatures as low as 20 to 40°C if the silica equilibrated with the same seawater $^{18}O/^{16}O$ ratio as today (with icecaps melted in) (Das Sharma et al. 1994). However, the age of these cherts is not well documented. Described as oolitic, they may be composed of silica deposited at more recent times. Alternatively, the formation water in equilibrium with depositing silica may have been enriched in ^{18}O as a result of deposition in a lagoon or brine lake, as is the case in contemporary settings (Gat 1996).

Sedimentologic Evidence

Archean first cycle clastic sediments, derived from exposed igneous or metamorphic rocks, are deeply weathered, with arkoses (feldspar-rich sandstones) only appearing in abundance in the Proterozoic (Lowe 1994), consistent with high Archean chemical weathering intensities preempting the survival of feldspar as a component of sediment. High Archean chemical weathering intensities plausibly resulted from high surface temperatures and atmospheric pCO_2 levels balancing a higher volcanic/metamorphic CO_2 source on somewhat smaller continental areas relative to more recent geologic time (the significance of weathering intensity is discussed more fully in chapter 8). Other sedimentologic evidence of high-intensity weathering in the Archean includes the great enrichment in chert and quartz in sediments of that age, abundant evaporites in early greenstones (Lowe 1994), and early and pervasive silicification (Knauth and Lowe 1978; Lowe and Knauth 1977). Similarly, a study of the geochemistry of 3 Ga shales from Zimbabwe strongly supports an inference of intense chemical weathering at the source of these sediments (Fedo et al. 1996). The evidence for a strongly stratified Archean ocean is consistent with high surface temperatures (Lowe 1994). Evidence for a much higher weathering intensity than now is also evident in sediments deposited at 1.9 Ga (Hayashi et al. 1997).

Sedimentary Gypsum Is Not a Good Temperature Constraint

An upper limit on surface temperature has been commonly cited, this for 3.5 Ga, from evidence for primary evaporitic gypsum precipitation (Walker 1982; Crowley and North 1991), providing a limit of 58°C because anhydrite is stable above that temperature in fresh water; in seawater the temperature is still lower (figure 7-6). However, this constraint appears tenuous given the metastable precipitation of gypsum in nature and in laboratory experiments, far above its stability field in place of anhydrite, even at 80°C (Berner 1971, Gunatilaka 1990, Cody 1976; Cody and Cody 1988). Metastable pre-

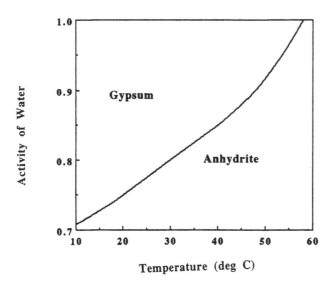

FIGURE 7-6.
The stability fields of gypsum ($CaSO_4.2H_2O$) and anhydrite ($CaSO_4$) (1 atm total pressure).

cipitation of gypsum above its stability field also may undercut the claim that evidence of possible coexisting evaporitic gypsum and halite in the Proterozoic gives an upper temperature limit of 18°C (Walker 1982). When anhydrite does form from a gypsum precursor, it often inherits the gypsum crystal form, the very evidence (in the form of pseudomorphs or molds) that is cited for the 58 and 18°C temperature limits (Gunatilaka 1990).

Lack of Fractionation of Sulfur Isotopes

There is a comparative lack of fractionation of sulphur isotopes between coexisting sulphide and sulphates in Archean and early Proterozoic sediments relative to Phanerozoic sediments (figure 7-7) (Ohmoto and Felder 1987; Knoll 1990; Bottomley et al. 1992). Knoll (1990) reviews possible explanations for this lack of fractionation in sediments dated at 3.5 Ga, concluding:

> . . . perhaps almost all sulphate in pore fluids was reduced biologically to sulphide in an essentially closed system with little fractionation because of high ambient temperatures (70°C or more)—a theory for which the geological record provides little supporting evidence. A generally acceptable solution to this problem has not yet been proposed (p. 12).

We cite here and in other published work just such supporting evidence.

Constraints on pH/pCO$_2$ Ocean and Atmospheric pCO$_2$

Atmospheric levels of pCO$_2$ on the order of 1 bar correspond to temperatures ranging from 50 to 70°C in the Archean/early Proterozoic (Schwartzman et al. 1993; Schwartzman and Shore 1996). This level is compatible with constraints on oceanic pH/pCO$_2$ (Grotzinger and Kasting 1993; see figure 7-8).

As an alternative to very high atmospheric pCO$_2$ levels, a methane and/ or ammonia greenhouse (Sagan 1977; Kasting et al. 1983; Zahnle 1986; Kasting and Grinspoon 1990; Sagan and Chyba 1997; Kasting 1997) has been proposed for the Archean, implying that somewhat lower atmospheric carbon dioxide levels achieved steady state by the weathering effect of high

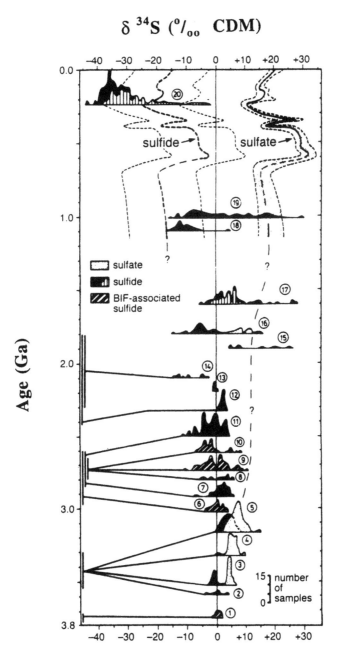

$\delta^{34}S\ (^o/_{oo}\ CDM)$

FIGURE 7-7.
Isotopic composition of sedimentary sulfide and sulfate as a function of age. (After Veizer 1994.)

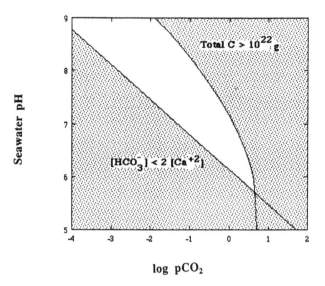

FIGURE 7-8.

Relationship between atmospheric pCO_2 and pH of seawater. (After Grotzinger and Kasting, 1993.) Possible composition of Archean and early Proterozoic ($>$1.6 Ga) seawater is shown in white.

surface temperature alone, or partially supplemented by pCO_2. However, there are indications that an antigreenhouse effect, arising from organic haze shielding, may reduce the viability of a methane-ammonia greenhouse (McKay et al. 1999). The plausibility of a methane greenhouse effect during the Archean may be diminished if an additional sink with respect to the atmosphere aside from photochemical dissociation of methane (Kasting et al. 1983) were present, thereby lowering steady-state methane levels. Perhaps both aerobic and anaerobic (utilizing sulfate reduction) methane-oxidizing microbes provided such a sink (Large 1983; Strauss et al. 1992a; Hinrichs et al. 1999). Aerobic microenvironments probably existed in the Archean, at least by 3.5 Ga if cyanobacteria did indeed emerge then, as recent evidence supports. The case for an ammonia greenhouse may be weakened if the pH of the Archean ocean was lower than the required minimum for the required production rate (pH 7.3; see Summers and Chang 1993). High pCO_2 during the Archean is consistent with just such moderately low pH model oceans (Grotzinger and Kasting 1993). If post-Huronian temperatures in the early Proterozoic were high, a reduced gas (e.g., methane) greenhouse is unlikely given the rise of atmospheric oxygen during this time, thus requiring high atmospheric pCO_2 levels.

Finally, a high pCO_2 pressure-cooker atmosphere at 3.8 Ga, inherited from the late bombardment, at steady-state with modest continents requires the present biotic enhancement of weathering factor to be on the order of 100 or more (see chapter 8 Appendix). A steady-state CO_2 greenhouse in the Archean, with the proposed high temperatures, provides a solution to the faint young sun paradox that is consistent with our best estimates of continental area and volcanic outgassing rates, without the need for other greenhouse gases. Other arguments for requiring a reducing atmosphere (i.e., necessary condition for origin of life, mechanism for liquid water on early Mars) have plausible alternatives.

A recent ingenious attempt to infer atmospheric pCO_2 levels prior to 2.2 Ga from the mineralogy of paleosols has derived an upper limit of $10^{-1.4}$ atm (Rye et al. 1995) from the absence of the iron carbonate siderite from these rocks. First, as Kasting (1997) has noted, an argument based on absence of a phase is a rather weak form of evidence. This inferred limit on atmospheric pCO_2 may be tenuous given the metamorphic/metasomatic alteration of the original paleosol and questionable assumptions (such as surface temperature and relevant phase equilibria; e.g., an atmospheric pCO_2 on the order of 1 bar could actually prevent, by lowering the pH of water involved in weathering, the formation of the phase, siderite, the absence of which is taken as the basis of the upper limit on pCO_2). An alternative ferrous iron–bearing phase formed during the original weathering might be "green rust," an iron hydroxide stable at low pO_2 and acid conditions (pH > 4) (Myneni et al. 1997). In contrast, for Archean and early Proterozoic banded iron formations (BIFs), siderite is commonly a primary phase, stable at the same high pCO_2 (Klein and Bricker 1977), consistent with seawater pH buffered above 6 (Grotzinger and Kasting 1993). Moreover, the coexistence of primary siderite and primary Fe silicates in these rocks apparently constrains pCO_2 at about 1 bar (Klein and Bricker 1977). There is intriguing evidence of present-day bacterial biomineralization of Fe hydroxide and silica in hot springs, a plausible model of some Precambrian BIFs (Konhauser and Ferris 1996), a possibility also suggested on the basis of studies of magnetite formation by thermophilic iron-reducing bacteria (Zhang et al. 1997).

Cyanobacterial CO_2 Concentrating Mechanism

The ability of cyanobacteria to increase their internal pCO_2 by up to 1000 times is consistent with an adaptation to declining CO_2/O_2 in the external environment (Strauss et al. 1992b; Badger and Andrews 1987) and therefore much higher levels of CO_2 in the Archean atmosphere, when cyanobacteria first emerged. The evolution of the CO_2 concentrating mechanism allowed photoautotrophs to maintain high levels of carbon fixation. The combined effect of declining CO_2 and increasing O_2 on the the efficient operation of Rubisco, the CO_2-fixing enzyme, have apparently led to the evolution of a CO_2 pump which greatly increases the internal pCO_2 in cyanobacterial cells (see Volk 1998, for a lucid discussion of Rubisco and the evolution of photosynthesis).

But What About the Huronian Glaciation at 2.3 Ga?

The Huronian glaciation is generally accepted by geologists as the earliest documented glaciation, although Young et al. (1998) have reported evidence for one still older, at about 2.9 Ga. Presumed tillite formations containing clasts with striations and polished facets dated at 2.2 to 2.3 Ga occur in Canada (Huronian formation), South Africa, and a few other locations (Harland 1983; Evans et al. 1997). Other purported evidence for a glacial origin of these deposits include the presence of dropstones, i.e., out-of-place rock fragments in finer sediment, usually interpreted as a result of ice-rafted sediment falling into marine deposits, striated basement, varve-like deposits and ice-wedge casts (Young 1991, 1993; Young and Long 1976). Similar evidence, particularly striated clasts and dropstones, has been reported for the purported glaciation at about 2.9 Ga (Young et al. 1998). However, all this evidence for glacial origin may have plausible alternative explanations. The Neoproterozoic (0.6–1 Ga) glaciations curiously follow more than a billion years later.

In what may prove to be a dramatic example of multiple discovery with major implications to the earth sciences and biology, Rampino (1992), Marshall and Oberbeck (1992), and Oberbeck et al. (1993) have proposed that some if not all ancient glacial deposits are really impact deposits, the ejecta of large impacts. If verified, a key upper temperature constraint for that time in the Proterozoic would be removed. The cited evidence includes the great volume of the deposits, thought inconsistent with a glaciomarine origin, and their incongruous association with indicators of warm climates. Evidently, even the classic evidence of glaciation, the polished and striated bedrock surfaces, can be produced by debris flows triggered by large impacts because these features are associated with debris flows of nonglacial origin. Oberbeck et al. (1993) have calculated the areal extents and different depths of impact debris predicted by the impact rate over the past 2 billion years, and this curve reproduces the global area-depth data from the presumed glacial deposits (figure 7-9). The possibility of late Precambrian (Neoproterozoic) glaciations is consistent with, but not required by, the proposed temperature history discussed here because temperatures drop significantly by 1.2 Ga, plausibly in the range that would permit glaciation. At least three possible explanations of the Huronian event are consistent with the above high temperature scenario (Rampino et al. 1996).

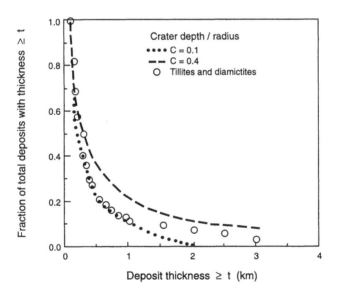

FIGURE 7-9.
Comparison of relative thickness distribution of tillites/diamictites with the relative thickness distributions of impact crater deposits from craters with depth/radius ratios C = 0.1–0.4 formed during the last 2 billion years. (After Oberbeck et al. 1993.)

FIGURE 7-10.
Glacial-like striations, faceting and polishing on pebbles from the Cretaceous-Tertiary
(K-T) impact deposit, Belize. Photo courtesy of Michael Rampino.

1. A large global temperature excursion (drop of 50°C) occurred, perhaps
triggered by an albedo increase associated with impact-generated debris.
Such an excursion was originally suggested by Knauth and Epstein (1976).
An oblique impact might have triggered just this excursion by creating an

orbiting ring around Earth, around long enough to reduce solar radiation reaching the atmosphere and surface (Schultz and Gault 1990). Just this mechanism has been suggested as the trigger for global cooling in the late Devonian (McGhee 1996). Reflective CO_2 ice clouds might prolong low surface temperatures (Caldeira and Kasting 1992a, but see Forget and Pierre-humbert 1997, who argue for a warming effect of CO_2 ice clouds).

2. Glaciation did not occur; the Huronian "tillites" are simply normal debris flows or are really the result of impact, which can generate features commonly identified as glacial (e.g., striations, polishing of clasts) which are found in the Cretaceous-Tertiary and other known impact deposits (Rampino et al. 1997; see figure 7-10).

3. A methane-dominated greenhouse from 2.8 to 2.3 Ga was destroyed by the rise of atmospheric oxygen, triggering glaciation (see Hayes 1994, for evidence of methane production). It then took some 10^8 years for volcani-

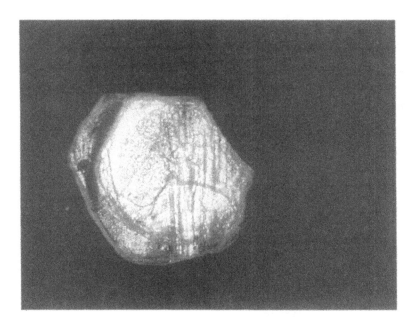

FIGURE 7-11.
Diamond (grain size 300 microns) from Popigai impact deposit. Diamonds were produced during impact 36 million years ago in northern Siberia. (Photo courtesy of Christian Koeberl; see Koeberl et al. 1997.)

cally derived CO_2 to build up in the atmosphere, returning surface temperatures to 50 to 60°C, consistent with the oxygen isotope record.

The Huronian deposits should be reinvestigated with these possibilities in mind. Evidence for impact would include shock features in quartz, occurrence of diamonds (figure 7-11), and other sedimentologic characteristics consistent with impact but not glacial origin. Paleotemperatures might be determined on unmetamorphed chert or carbonates of Huronian age (the only published data are on metamorphosed carbonates giving altered light oxygen isotopic ratios; see Veizer et al. 1992).

The Dearth of Obligate Psychrophiles Today

Psychrophiles are organisms with an optimal temperature for growth of 15°C, with a maximum growth temperature below 20°C, and a minimal temperature of growth of 0°C or lower (Brock and Madigan 1991). Knoll and Bauld (1989) observed that if the apparent small number of species of obligate psychrophiles is real, it "could reflect the course of Earth's history (i.e., the relatively late appearance and discontinuous presence in time of extremely cold environments)," an explanation consistent with temperature trajectory H in figure 7-2.

8 DID SURFACE TEMPERATURES CONSTRAIN MICROBIAL EVOLUTION?

An evolutionary puzzle; oxygen as a constraint on evolution? The postulated link between temperature and major evolutionary developments in the Precambrian. The implications of a much warmer Precambrian Earth surface to evolutionary biology. Microbial phylogeny and the upper temperature limit for growth. Altitudinal, latitudinal, and diurnal temperature variations on a very warm Archean/early Proterozoic Earth surface. Implications to the history of biotic enhancement of weathering. A cloud-free Archean: explanation for high surface temperatures? Biospheric evolution and weathering. A geophysiological model for the evolution of the biosphere. From "hothouse" to "icehouse": the increase in diversity of habitats for life over geologic time.

Appendix: Climate model description. Surface temperatures on Earth since the early Archean. The carbonate-silicate geochemical cycle at 3.8 Ga, for an abiotic Earth surface just at the transition to biotic colonization.

An Evolutionary Puzzle: Oxygen as a Constraint on Evolution?

Maynard Smith and Szathmary (1995) remarked:

> Astonishingly, the time that was needed to pass from inanimate matter to life is four times shorter than needed for passing from prokaryotes to eukaryotes [this is likely a very conservative estimate] . . . it is hard to argue that they [the steps to form eukaryotes] are more difficult than to establish a genetic code.

They then went on to propose an explanation based on existing competition of procaryotes (p. 145). Furthermore, they were also puzzled about the timing of the subsequent emergence of Metazoa, but offer the conventional explanation, the rise of atmospheric oxygen. Let us examine this critical problem of the timing of emergence more closely.

Cloud (1976) proposed that atmospheric pO_2 levels determined the timing of events in biotic evolution (figure 8-1). Aerobic microenvironments in the Archean apparently preceded the emergence of aerobic eucaryotes, which in turn preceded their first likely presence in the fossil record. Indeed, it has been suggested recently that aerobic respiration emerged very early (Castresana and Saraste 1995). Some deeply rooted hyperthermophiles (figure 8-2) are microaerophilic (e.g., Aquifex, a Eubacteria), suggesting either that traces of free oxygen were present very early on (Towe 1994) or that this metabolism, a chemoautotrophic oxidation of hydrogen, was acquired

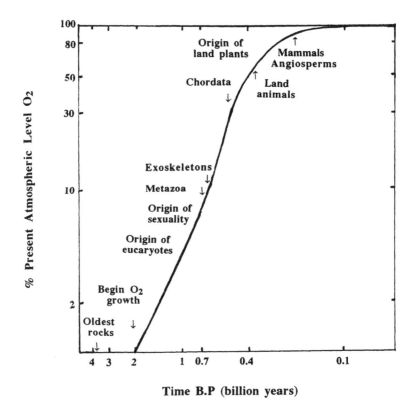

FIGURE 8-1.

Apparent timing of events in biospheric evolution compared with hypothetical levels of atmospheric oxygen. (After Cloud 1976.)

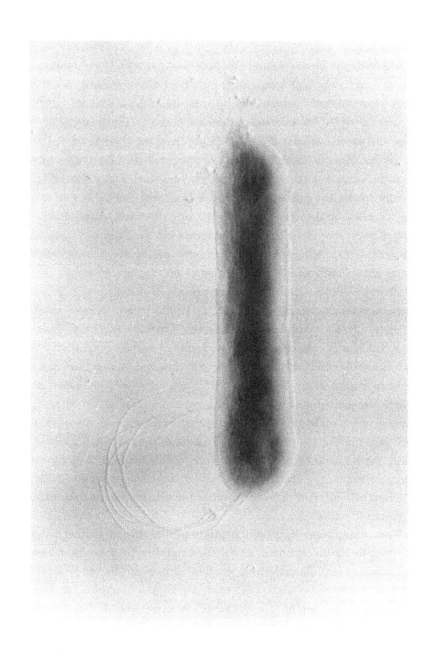

FIGURE 8-2.

Aquifex pyrophilus, a marine hyperthermophilic microaerophile, length 2.5 microns, with an optimal growth temperature of 85°C (upper limit 95°C). (Courtesy of K. O. Stetter.)

later in Earth history (Dyer and Obar 1994). Low levels of free oxygen preceding oxygenic photosynthesis may have been generated by photodissociation of water, or by high temperature pyrolysis close to erupting lava. If oxygenic photosynthesis occurred by no later than 3.5 Ga (Schopf 1992) the occurrence of aerobic microenvironments, with pO_2 on the order of 1 bar (e.g., in cyanobacterial mats) plausibly predated the emergence of aerobic eucaryotes. Knauth (1998) proposed that prior to about 2 Ga oxygen solubility in the ocean was lower than in later times because of the combined effect of high salinity and high temperature. He suggested that the lack of oxygen may have held up the emergence of metazoans. However, his contention that early oceans were highly saline is contingent on the absence of continental platforms (sites of salt deposition), a debatable assumption (see discussion on continental growth rates in the Appendix to this chapter). Moreover, aerobic conditions in the ocean should still have occurred in regions of high cyanobacterial productivity prior to 2 Ga. Indeed, the carbon isotopic record of kerogens apparently supports this conclusion (Strauss et al. 1992a). Thus, the lower limits of pO_2 in surface microenvironments for the emergence of aerobic eucaryotes and Metazoa may have be reached earlier than their upper temperature limits (60°C and 50°C, respectively). The rise of atmospheric oxygen by 1.9 Ga (15% PAL according to Holland 1994) predated the emergence of Metazoa, which may require less than approximately 2% PAL (e.g., mud-dwelling nematodes; see Runnegar 1991). On the other hand, the rise of atmospheric oxygen may well have constrained the emergence of megascopic eucaryotes, particularly Metazoa, as originally argued by Cloud (1976), with the explanation being the diffusion barrier of larger organisms (Raff 1996).

Thus, why did aerobic environments plausibly sufficient for their metabolism predate the emergence of aerobic eucaryotes and Metazoa by at least 0.5 to 1 billion years? The following possibilities may explain this observation:

1. The earliest aerobic (mitochondrial) eucaryote and Metazoan fossils were not recognized or not preserved.

2. Physical constraints other than oxygen level prevented their emergence; temperature is one likely constraint.

3. Evolution is that slow (probably the least likely explanation, given the rapid evolutionary potential of at least procaryotes as shown in laboratory experiments).

If surface temperature was the critical constraint on microbial evolution as we have suggested (Schwartzman et al. 1993), then the approximate upper temperature limit for viable growth of a microbial group should equal the actual surface temperature at the time of emergence, assuming that an ancient and necessary biochemical character determines the presently determined upper temperature limit of each group. The latter assumption is supported by extensive data base of living thermophilic organisms. No cyanobacteria have been found to grow above about 70°C, despite an apparent 3.5 billion-year ancestry of oxygenic photosynthesis. Similarly, eucaryotes have an upper limit of 60°C and have had 2 billion years to adapt to life at higher temperatures.

The upper temperature limit for viable growth has been determined for the main organismal groups (table 8-1). This limit is apparently determined by the thermolability of biomolecules (e.g., nucleic acids), organellar mem-

TABLE 8-1.

Upper Temperature Limits for Growth of Living Organisms and Approximate Times of their Emergence

Group	Approximate Upper Temperature Limit (°C)	Time of Emergence (Ga)
Plants		
Vascular plants	45	0.4[a]
Mosses	50	0.5[a,b]
Metazoa	50	1–1.5[c]
Aerobic Eucaryotes	60	2.6[a,c]
Procaryotic microbes		
Cyanobacteria	70–73	≥3.5[a]
Methanogens	>100	>3.8[d]
Extreme thermophiles	>100	>3.8[d]

Temperatures from Brock and Madigan (1991).
[a]Fossil evidence; for earliest eucaryotes, 2.1 Ga (Han and Runnegar 1992).
[b]Primitive bryophytes.
[c]Problematic fossil evidence for Metazoa, molecular phylogeny.
[d]Molecular phylogeny.

branes, and enzyme systems (e.g., heat shock proteins) (Brock and Madigan 1991). For example, the mitochondrial membrane is particularly thermolabile, apparently resulting in an upper temperature limit of 60°C for aerobic eucaryotes. Could the upper temperature limit of 50°C for Metazoa be linked to the thermolability of proteins essential to blastula formation, or of the synthesis of collagen, an essential structural protein? Clearly, fundamental research is needed to better understand the biochemical and biophysical basis for the upper temperature limits of normal metabolism of the organismal groups.

Anaerobic eucaryotes may have preceded aerobic eucaryotes before the endosymbiogenic event creating mitochondria (Sogin et al. 1989), although recent data suggest that the nucleus and mitochondrion arose nearly simultaneously in the eucaryote cell (Sogin 1997; Gray et al. 1999). A higher temperature limit for ancestral amitochondrial than mitochondrial eucaryotes would be consistent with suggestions that mitochondria are more thermolabile than nuclei. It would be interesting to determine whether any living anaerobic eucaryotes are found to be viable above the apparent limit for aerobic eucaryotes (about 60°C).

We have proposed that surface temperatures were the critical constraint on microbial evolution, determining the timing of major innovation (Schwartzman et al. 1993). Surface temperatures at emergence corresponded to the upper temperature limits for each group (i.e., cyanobacteria 70°C, at 3.5 Ga; aerobic eucaryotes 60°C, at 2.6 Ga; and Metazoa 50°C, at 1.0–1.5 Ga), as shown in figure 8-3 and table 8-1. This temperature history is consistent with inferred climatic paleotemperatures from the oxygen isotopic record of Precambrian cherts and carbonates (see chapter 7).

Fungi now have thermophilic species that can grow to an upper limit of about 62°C (Brock 1978). Because the divergence of fungi and Metazoa, sister kingdoms, apparently occurred between 1 and 1.5 Ga (Chapman 1992), it is not clear whether fungi modestly preceded Metazoa, a scenario consistent with fungi's higher upper temperature limit, or that thermophilic fungi adapted more recently to living in a higher temperature environment. Similarly, does the fact that the more deeply rooted bryophytes have a 5°C higher upper temperature limit than do vascular plants indicate an ancient temperature constraint on the order of emergence of these two groups among the Plantae? If the geologic evidence for the occurrence of ice ages is accepted for the late Proterozoic (but see their reinterpretation as possible

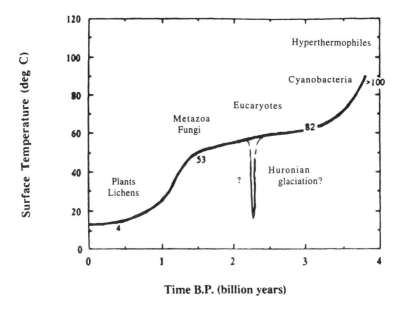

FIGURE 8-3.
Proposed surface temperature history of Earth, with key evolutionary developments noted. Numbers near curves are the ratios of the present biotic enhancement of weathering to the value at that time (*BR*) computed for preferred model b.

impacts in Oberbeck et al. 1993 and Rampino 1994) then mean global temperatures reached some 10 to 20°C long before the fossil evidence for Plantae. Temperatures of 25 to 43°C near sea level are also indicated from the chert oxygen isotope record (Kenny and Knauth 1992). Thus, the possibility of earlier emergence of Plantae is problematical but deserves closer attention. An interesting possibility is that each major innovation first occurred in cooler environments than found in the ocean. The only such plausible environment cooler than at sea level at the high surface temperatures favored here for the Archean/early Proterozoic appears to be mountain lakes, whose sediments and fossils are unlikely to be preserved in the geologic record.

In the scenario outlined above, the genetic potentiality for rapid evolution is simply realized as soon as (relative to a geologic time scale, of course) external conditions allow its expression. If confirmed, this conclusion would have major implications to evolutionary biology and our understanding of the evolution of the biosphere. This explanation of evolutionary emergence is more deterministic than that given by Maynard Smith and Szathmary (1996):

"To explain a particular event, say the origin of eucaryotes, is to show that, given plausible initial conditions, the event, if not inevitable, was at least reasonably likely."

Altitudinal, Latitudinal, and Diurnal Temperature Variations on a Very Warm Archean/Early Proterozoic Earth Surface and Possible Implications to History of Weathering

The lapse rate, the change in air temperature with elevation, will determine the air temperature on mountains (as previously mentioned, the ground temperature may be considerably higher, depending on the local surface albedo). If we assume that the maximum height of mountains is 10 km (Everest is 9.1 km), then for an average lapse rate of 6.5°C/km (Henderson-Sellers and McGuffie 1987) the lowest temperature on the Earth's surface for a sea level temperature of 70°C is 5°C; for 50°C, −15°C. Note that the lapse rate varies from 4°C/km (moist) to 10°C/km (dry).

Latitudinal differences in temperature decrease as global mean temperature increases, judging from the paleoclimate record of the Cenozoic (Hoffert and Covey 1992). While the processes that equalize temperature are not well understood, latitudinal differences in temperature are predicted to vanish for a global mean temperature of ≥30°C (Hoffert, personal communication). Likewise, diurnal variations would be expected to significantly diminish as the mean temperature increases, as a result of greater heat storage in the ground and troposphere, which is radiated at night.

An Archean/early Proterozoic Earth likely rotated at a faster rate than now, with a day some 10 hours shorter, based on the progressive slowing of the rotation rate from lunar tides (Walker and Zahnle 1986; Zahnle and Walker 1987). Based on a computer simulation of the Precambrian Earth, taking into account the effects of a faster rotating Earth, and assuming a near present day temperature, Jenkins et al. (1993) concluded there would have been a modest temperature elevation of a few degrees from this effect, as well as an increase in the latitudinal temperature gradient from the weakening of the tropical Hadley circulation, which carries heat from the equator poleward. However, much higher mean temperatures would have likely obliterated the latter tendency.

If the onset of latitudinal differences in temperature occurs at or below a mean global temperature of about 30°C, then the preferred temperature history for the Earth (figure 8-3) predicts the commencement of seasonality, as well as bigger fluctuations in diurnal temperature starting at about 1.5 Ga. Furthermore, as temperatures decrease, an expansion of mountain area subject to freeze/thaw cycling is expected. All of these factors may have contributed to a significant increase in physical weathering via ice wedging and thermal cracking (insolation weathering), generating silt size mineral and rock particles (Moss et al. 1981; Dove 1995; Fahey 1983; Summerfield 1991). The production and delivery of chemically immature mineral and rock particles to lowland flood plains might well have significantly increased the chemical weathering rate (Edmond 1992); thus, if confirmed, the long-term biotically mediated cooling in the Archean/Proterozoic set up the conditions for the onset of frost wedging as an important mechanism of physical weathering, itself leading to further cooling by its synergy with chemical weathering. The proposed influence of the onset of seasonality and intensified frost wedging in mountains at a given surface temperature and steady state chemical weathering flux (at T_i, W_i) on chemical weathering intensity ($W = (V/V_o)(A_o/A)$, i.e., the silicate C sink (flux/unit land area) is shown in figure 8-4. Note that a transient increase in W is expected, which since the Archean would not last more than 10^7 years, given the likely pCO_2 level in the atmosphere/ocean system (Sunquist 1991). A new steady-state pCO_2 and temperature (T_j) is reached at W_i, if A and V do not change significantly in this interval. A similar pattern might be expected with sudden events in biotic evolution, such as the emergence of higher plants, but their spread over the land surface probably took more than 10^7 years.

If the weathering intensity W is plotted versus surface temperature, then an activation energy can be computed for the temperature range of 70 to 60°C (corresponding to a range of about 3.5–2.6 Ga). A model of (V and A) variation and a ratio of biotic enhancement of weathering factors for the two times (i.e., $B_{3.5\ Ga}/B_{2.6\ Ga}$) is assumed (figure 8-5; note that the temperature decrease from B to A [now], with a relatively modest decrease in weathering intensity, has been previously explained as a consequence of the increase in biotic enhancement of weathering). A marked decrease in W from 3.5 to 2.6 Ga is consistent with sedimentologic evidence already cited. For the preferred model "b" (see Appendix for definition), with $B_{3.5\ Ga}/B_{2.6\ Ga} = 0.5$ (note that precipitation is assumed to saturate at about 55°C), the computed $E = 58$ kJ/mole, similar to the watershed value of White and Blum (1995).

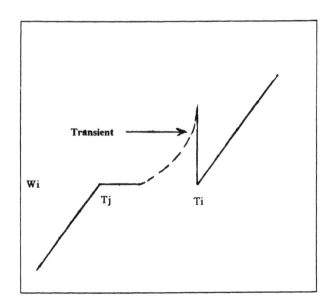

Surface Temperature

FIGURE 8-4.
The effect of decreasing global surface temperature on the chemical weathering intensity (W). Frost wedging/climatic seasonality occurs at T_j, generating a transient change in W with steady state reestablished at W_i.

I claim here no uniqueness in the model result, only that plausible activation energies of weathering can be computed. Obviously, weathering intensities and other factors need to be better established. One possible approach to inferring weathering intensities might be via studies of clastic sediments. For example, Heins (1995) has claimed that the stability of mineral interfaces in clastic sediments is sensitive to temperature and precipitation.

Implications to the History of Biotic Enhancement of Weathering

The inversion of the atmospheric pCO_2 history of the Phanerozoic in the GEOCARB model (Berner 1993, 1994) gives an estimate of the progressive increase in biotic enhancement of weathering as a result of the emergence

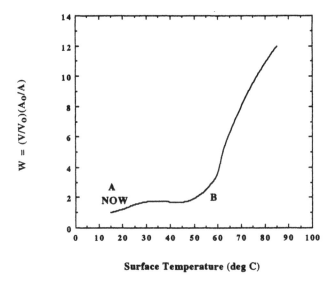

FIGURE 8-5.
Weathering intensity W (steady state) as a function of surface temperature. Model b is
assumed for variation of W with age.

and spread of vascular plants (see chapter 5). We have attempted the same
approach for the Precambrian (Schwartzman and Volk 1991a; Schwartzman
et al. 1993; Schwartzman and McMenamin 1993; Schwartzman and Shore
1996). Our first attempt (Schwartzman and Volk 1991a) assumed the conventional scenario ("C") for the temperature history, whereas the more recent attempt assumed a much warmer Precambrian ("H") (figure 8-6). Scenario H is identical to the trajectory in figure 8-3, consistent with the hypothesis that surface temperatures have been a critical constraint on microbial
evolution, determining the timing of major innovations and with the paleotemperature isotopic record of cherts and carbonates (see review by Knauth
1992). There are of course perturbations in the Phanerozoic; scenario H is a
first approximation, ignoring these deviations from the trend. Also shown in
figure 8-6 is a large excursion of temperature (E–F), that would be required
by a Huronian glaciation at about 2.3 Ga. This excursion is of course not necessary if the Huronian "tillites" are not really of glacial origin (see chapter 7).

Let B_R be the ratio of the present biotic enhancement of weathering to
that at a time in the past. We will show below that scenario H probably
requires $B_R \geq 80$, just using the inversion of inferred temperatures prior to

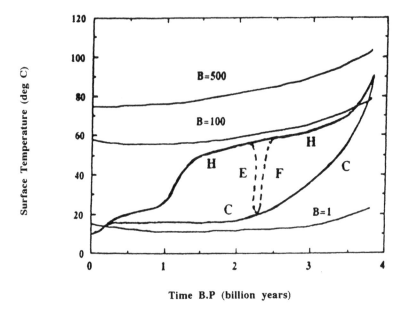

FIGURE 8-6.

Earth surface temperature versus Time B.P. (billion years). Two alternative scenarios are noted (H and C, with possible excursion E-F). Temperature trajectories for $B = 500$, 100, and 1 are also indicated (for preferred model b) (B is B_R in text).

the Huronian event, provided carbon dioxide and water vapor were the sole greenhouses gases (Schwartzman et al. 1993 and Schwartzman and McMenamin 1993 give a limit of about 50, using a different greenhouse function than here). If post-Huronian temperatures in the early Proterozoic follow scenario H, as supported by both chert and carbonate paleotemperature data, then this lower limit on B is even more robust because a reduced gas (e.g., methane) greenhouse is unlikely with the rise of atmospheric oxygen during this time. This limit on B_R is a minimum for biotic enhancement over abiotic conditions because some biotic acceleration of weathering is even plausible in the Archean, with partial thermophilic colonization of the land (Schwartzman and Volk 1989, 1991a). Several abiotic factors, namely impact-derived soil (regolith) and greater volcanic activity (volcanics weather faster than other silicate rocks) would by themselves have favored a higher weathering intensity, relative to the present, and have tended to pull temperatures down, strengthening the requirement for a high present biotic enhancement of weathering to keep temperatures high in the Archean.

The modeling for the above results is similar to the approach in Schwartz-man and Volk (1991a). However, a modified greenhouse function was used for atmospheric $pCO_2 > 0.03$ bar (see Appendix for detailed model description and related issues). Carver and Vardavas (1994, 1995) have similarly modeled the long-term carbonate-silicate cycle, with an assumed temperature history close to that assumed in Schwartzman and Volk (1991a). We have since changed our mind on the probable Archean to mid-Proterozoic temperature trajectory.

Also plotted in figure 8-6 are the temperature curves for B_R values of 500, 100, and 1, to illustrate the effect of varying the biotic enhancement ratio. These trajectories were computed from our model of the steady-state atmospheric pCO_2 determined by the balancing of the volcanic source and weathering sink. These curves for different values of B_R give temperatures for both past biotically enhanced surfaces and abiotic surfaces with the same value of B_R since the equations for both cases are identical in form (see next section).

For $B_R > 160$, the surface temperature remains above 60°C on an abiotic Earth; that is, such an Earth is habitable for thermophiles but not complex life (similar to the conclusion we reached in Schwartzman and Volk 1989; see discussion in chapter 10).

A Cloud-free Archean: Explanation for High Surface Temperatures?

Returning to an earlier discussion of the DMS/CCN story (see chapter 1), if indeed there was much lower production of CCNs in the Archean, as a result of the absence of DMS production by eucaryotic algae, with no other source comparable with today's, the surface albedo could have been as low as 0.1. A naive calculation of this effect on surface temperature gives increases on the order of 30°C. However, taking into account the new steady-state in the silicate-carbonate cycle brings this increase down to a more modest increment of 5 to 10°C relative to the constant surface albedo case for the Archean (Volk 1998 computed a similar result). Given the likelihood that other CCN sources may have been more important in the Archean (e.g., from stratospheric oxidation of H_2S), the tentative conclusion is that albedo variations are unlikely to explain the high apparent surface temperatures in the first two thirds of Earth history.

Biospheric Evolution and Weathering

A high carbon dioxide "pressure-cooker" atmosphere, inherited from the vigorous degassing and the late bombardment of cometary/meteoritic impacts on the early Earth, is assumed just prior to the initial biotic colonization of land, with surface temperatures 70 to 85°C or higher (Kasting and Ackerman 1986, Chyba et al. 1990; Delsemme 1992; Owen and Bar-Nun 1995).

The late bombardment ended at 3.75 to 3.80 Ga, with the last impact having energy sufficient to sterilize the Earth surface (Oberbeck and Mancinelli 1994). The origin of life may well have occurred earlier (4.2 Ga), with surface life continually destroyed at big impact times. Gogarten-Boekels et al. (1995) outlined an interesting scenario that takes into account the latest molecular phylogenetic evidence for the last common ancestor of the Archea, Eubacteria, and Eukarya being a full-fledged procaryotic cell. The hyperthermophiles (optimal temperatures for growth >85°C), deeply rooted in the phylogenetic tree, all apparently require the activity of the reverse gyrase enzyme (Bouthier de la Tour et al. 1990; Forterre et al. 1995; Forterre 1995a, 1995b, 1995c), which is inferred to have resulted from gene fusion in a lower temperature procaryote, which is termed *mesophilic*. Hence, the last common ancestor of the three domains is believed to have been mesophilic, with hyperthermophily resulting from the mesophile adapting to higher temperatures in hydrothermal reservoirs in the crust. In this scenario (figure 8-7) only two branches, the hyperthermophilic Archea and Eubacteria, made it back into the ocean and land after the last sterilizing impact at about 3.8 Ga. However, at least one thermophile (near hyperthermophile), Thermus thermophilus, apparently lacks reverse gyrase activity, yet has an upper temperature limit of 85°C (Kristjansson and Stetter 1992), suggesting that this temperature is consistent with the environment for the progenitor of hyperthermophiles. A surface temperature of 85°C is compatible with a pressure-cooker atmosphere of some 10 bars of carbon dioxide at around 4 Ga. There is, however, an ongoing debate on the interpretation of the reverse gyrase story, as well as whether the last common ancestor was mesophilic, thermophilic, or hyperthermophilic. A recent paper argued that the LCA was likely not hyperthermophilic, based on the observed correlation of the contents of guanine and cytosine in the ribosomal RNA of thermophiles with their optimal growth temperatures (Galtier et al. 1999). However, a higher temperature for the LCA would be inferred from their data if the upper temperature

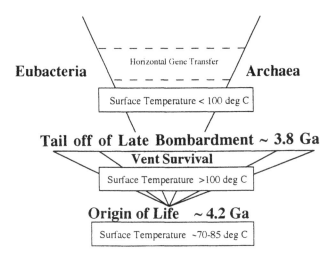

Eubacteria — Horizontal Gene Transfer — Archaea

Surface Temperature < 100 deg C

Tail off of Late Bombardment ~ 3.8 Ga

Vent Survival

Surface Temperature >100 deg C

Origin of Life ~ 4.2 Ga

Surface Temperature ~70-85 deg C

Origin of Earth 4.55 Ga

FIGURE 8-7.
Scenario for early biotic evolution. (After Gogarten-Boekels et al. 1995.)

limit for growth, a more plausible limit on emergence than the optimal temperature, is used. There are now over a dozen organisms whose genomes have been completely sequenced, with more to come (Doolittle 1998b). The results are now confounding molecular phylogenists (Pennisi 1998). Horizontal gene transfer between coexisting organisms appears now to have been the rule rather than the exception (Olendzanski and Gogarten 1998), which makes the interpretation of phylogenies from even complete genomes difficult. Nevertheless, the original hypothesis of Woese (1987) that early life was thermophilic or even hyperthermophilic (Pace 1997; Barns and Nierzwicki-Bauer 1997) is still robust (Gogarten, personal communication). Wachtershauser (1998) argues persuasively for a high temperature origin of life and by implication for a thermophilic LCA from the high improbability of enzymatic systems adapting from mesophily to hyperthermophily rather than the reverse path.

Surface cooling of the biosphere has been determined by its evolution, constrained by abiotic boundary conditions (i.e., luminosity of the Sun, continental area, outgassing rate). We proposed that soon after the origin of life this process commenced with the colonization of land by thermophiles/hy-

perthermophiles at around 3.8 Ga, resulting in the enhancement of weathering rates, and the sequestering of carbon dioxide into limestone deposits,
thereby cooling the Earth's surface (Schwartzman and Volk 1989, 1991a;
Oberbeck and Mancinelli 1994). Natural selection should have promoted
the survival of those mutants among land biota with greater nutrient (essential minor and trace elements) extraction ability and water retention growth
habits. The binding of mineral and rock particles by extracellular polymers
such as polysaccharides produced by procaryotic colonies is well known from
studies of natural consortia growing in and on the surface of soils (Campbell
1979). These characteristics are precisely those that would have increased the
weathering rate per land area, thus accelerating the removal of CO_2, bring
down global temperatures.

Could higher levels of ultraviolet (UV) radiation at the surface have prevented microbial colonization of land in the Archean? Other gases in the
atmosphere such as traces of sulfur dioxide could have acted as a UV screen
in the Archean (Kasting et al. 1989), in the absence of oxygen and hence an
ozone screen. Even without other UV screens, much higher levels of UV radiation at the surface could have been shielded by the organic products (e.g.,
mucus) of the bacterial colonies themselves (Lovelock 1988) and the stromatolitic growth habit (Margulis et al. 1976) (but for another view on this
issue see Towe 1994, who contends that minor levels of free oxygen provided just this UV screen). Pierson (1994) argued that an ozone screen was
likely a necessary condition for large, multicellular life on land. Thus, the
rise of atmospheric oxygen after about 2 Ga, and its concomitant strong UV
shield may be linked to a significant increase in the productivity of the land
biota, particularly for eucaryotic algae, and biotic enhancement of weathering, leading to the apparent profound cooling in the mid-Proterozoic.

Why did atmospheric oxygen rise in the Proterozoic if oxygenic photosynthesis began at least 1.3 billion years earlier? What was the "sink" for the
oxygen produced by cyanobacteria? The oxidation of dissolved ferrous iron
in the ocean was likely a major sink. Another likely contributing sink was
the reaction of reduced gases, such as hydrogen, from volcanic outgassing.
Kasting, Eggler, and Raeburn (1993) argued that the key factor in the rise
of oxygen was the progressive oxidation of the Archean mantle from the
subduction of water in the hydrated sea floor and the release of reduced volcanic gases. By about 2 Ga, the upper mantle was oxidized enough so the
oxidation state of volcanic gases increased, leading to a decrease in the consumption rate of atmospheric oxygen.

TABLE 8-2.
Evolution of Land Biota

Age (Ga)	Soil Community (New or Dominant Land Biota)
3.8-3.5	Hyperthermophiles, thermophiles (methanogens, thermoacidophiles, green nonsulphur bacteria, Thermus)
3.5-2.6	Cyanobacteria
2.6-2.3	Cyanobacteria mats with eucaryotes[a]
2.3-1.5	Eucaryotic/procaryotic mats[b]
1.5-0.65	Algal mats, primitive lichens (primitive fungi?), first Metazoa
0.65-0.4	Bryophytes[c] (the emergence of Hypersea; see McMenamin and McMenamin (1994)
0.4-0.2	Vascular plants, lichens[d]
0.2-0	Vascular plants (angiosperms, grasses)

[a]This likely community can be viewed as an anti-lichen by the reversal of the procaryote/eucaryote spatial relationship (i.e., procaryotic matrix with embedded eucaryotic cells, the inverse of lichen with cyanobacteria as the phycobiont). Perhaps modern microbial mats (see description of species diversity in Brown et al. 1985) are in some ways a model of ancient anti-lichens.
[b]For example, possibly the actinolichen, a symbiosis of actinobacteria and green algae. Actinolichens have been synthesized in the laboratory, and one example has been reported in nature (Hawksworth 1988).
[c]Earliest fossil: lower middle Ordovician (Strother et al. 1996).
[d]Earliest fossil of vascular plant, early Silurian (Cai et al. 1996); earliest fossil of lichen, early Devonian (Taylor et al. 1995).

Each new innovation in the microbial soil community resulted in greater biotic enhancement of weathering, culminating in the rhizosphere of higher plants (Schwartzman et al. 1993). The long-term increase of a major factor—soil stabilization—as biotic evolution changed the dominant land biota from procaryotes to vascular plants in the Phanerozoic has already been discussed. A somewhat speculative history is shown in table 8-2 (at this stage it must be speculative because the fossil record of Precambrian land biota is extremely sparse). The postulated evolution of land biota corresponds to the best current inferences from fossils and molecular phylogenetic reconstructions (see earlier discussion) that put hyperthermophiles at the base of the tree of life. Thus, the earliest land biota likely included hyperthermophilic methanogens, chemoautotrophs consuming atmospheric carbon dioxide in a pressure cooker atmosphere, along with traces of hydrogen produced by photodissociation of water. Atmospheric hydrogen levels of 1% for 10^8 years is apparently compatible with tropospheric photochemistry (Walker 1977). Other members of the earliest land biota probably included other hyperthermophiles such as the ancestors of filamentous *Chloroflexus aurantiacus* (a photoautotroph feeding on hydrogen, hydrogen sulfide, and carbon dioxide), now found in hot springs, as well as *Pyrodictum* and *Acidanus* (consumers of hydrogen and free sulfur). Other living models of this earliest land biota are found in the diverse community of strict anaerobes living in hot springs (submarine hydrothermal environments), often in matlike colonies on rock substrates. The postulated ancient communities covered the early Archean continents and volcanic islands with similar microbial mats, which dried out between rain events, only to be revived when wet.

Cyanobacteria, along with oxygenic photosynthesis, likely joined this land consortia not later than 3.5 Ga, judging from the inferred fossil record of the Warrawoona cherts (Schopf 1992, 1993). By providing aerobic microenvironments, these mats likely were home for the earliest aerobic eucaryotes, although their emergence was likely delayed by some billion years because of ambient temperatures above their upper limit of metabolism (60°C). These late Archean/early Proterozoic communities were plausibly antilichens, reversing the contemporary lichen spatial relationship between photoautotrophs and heterotrophs. Modern procaryotic mats similarly contain diverse eucaryotes such as diatoms (Brown et al. 1985).

By the mid-Proterozoic, with the rise of atmospheric oxygen, thicker, more productive eucaryotic mats likely dominated the land biota, with stronger soil-stabilizing power and water retention. By the late Proterozoic, new organisms such as primitive lichens likely emerged, with primitive fungi such as chytrids constituting the host for cyanobacterial or eucaryotic cells (modern lichens have with few exceptions *Ascomycetes* or *Basidiomycetes* host fungi). Even these primitive lichens may have been predated by actinolichens, having heterotrophic bacteria, actinobacteria, hosts; modern descendants have hyphae-like growth, forming matlike colonies. By the early Paleozoic, true plants emerged, first bryophytes.

The elevation of soil pCO_2 by prevascular plant land biota—first microbial, then bryophytic—was perhaps even comparable with later vascular plant–dominated soils, thereby contributing to chemical weathering (Keller and Wood 1993; Yapp and Poths 1994). The appearance of vascular plants in the early Silurian, with their global expansion by Carboniferous times, may have raised the soil pCO_2 even higher by root respiration and decay of organic matter. More important was the likely increase in soil thickness and stabilization relative to the earlier microbial and bryophyte land biota (Wright 1990). A new factor leading to an acceleration of chemical weathering and cooling arose, the rhizosphere (the association of mycorrhizae and plant root), with its multifold factors (organic acid and chelating agent production and increase in contact with soil minerals) (Berner 1995a; Retallack 1997). The emergence of a significant rhizosphere is documented from early Devonian fossils (Elick et al. 1998). Perhaps the great glaciation of the Carboniferous/Permian was related in part to this new factor (Algeo et al. 1995; Berner 1995a).

Much later, the emergence of angiosperms and their competition with conifers by early Tertiary may have further increased weathering rates, apparently because of the mycorrhizal relationships of angiosperms, which led to more efficient nutrient extraction from minerals, increasing the rate of CO_2 removal from the atmosphere (Volk 1989b). However, Robinson (1991) argued that the contemporary field data used by Volk to infer higher weathering rates for angiosperms was biased by more weatherable rocks in these watersheds. She concluded that there is no demonstrable difference between the chemical weathering rates of angiosperms and gymnosperms. Thus, it is problematic whether the increase of angiosperms contributed to a global cooling trend, which indeed is apparent in the last 100 million years. The later cooling may be more linked to a decrease in sea-floor generation rate (and outgassing). On the other hand, the emergence of widespread grasslands, with their deep root systems and sink for silica, could have promoted higher weathering rates and global cooling in the mid-Miocene (Johansson 1993, 1995).

The preferred scenario for surface temperature as a function of geologic age is shown in figure 8-3. Second-order perturbations arising from such factors as continental drift, episodic burial of organic carbon, and pulses of intense volcanism are ignored (particularly relevant in the Phanerozoic, when atmospheric pCO_2 levels decrease). Marked cooling in the mid-Proterozoic is indicated, consistent with the apparent emergence of Metazoa at 1.0 to 1.5 Ga (Chapman 1992; Morris 1993) and paleotemperatures of 25 to 43°C on 1.1 to 1.2 Ga cherts (Kenny and Knauth 1992). This cooling may have been the result of the rise of atmospheric oxygen, as already pointed out, and more extensive colonization of the land surface by algal mats (Beeunas and Knauth 1985; Horodyski and Knauth 1994). This increase in terrestrial biotic productivity, along with the onset of frost wedging in mountains, likely substantially increased the biotic enhancement of weathering. Note that although frost wedging is an abiotic process, here I argue that it commenced as an important physical weathering factor as a result of global biotically mediated cooling, thus bringing in this factor as a component of the cumulative global biotic enhancement of weathering. From the pattern of inferred variation of the U/Th ratio in the depleted mantle over geologic time, Collerson and Kamber (1999) suggest that the reinjection of oxidized uranium begins at about 2 Ga, accelerated by enhanced biotic weathering and continental erosion, consistent with the above argument.

Although the first definitive fossil evidence for Metazoa dates at 0.65 Ga (McMenamin and McMenamin 1990), an older problematic record does exist (Hofmann 1994; Robbins et al. 1985; Breyer et al. 1995; Fedonkin et al. 1994; Seilacher 1997; Seilacher et al. 1998). The story of the earliest Metazoa may be analogous to that of Eucarya (Sogin et al. 1989) in the unlikely preservation of microscopic soft-bodied organisms as fossils. Davidson et al. (1995) noted that "micrometazoan ancestors would not have left a fossil record because of their small and probable lack of skeletonization." These ancestors are proposed to "constitute a cryptic pre-Ediacaran record." The emergence of Metazoa at 1 Ga or earlier are supported by the results of a molecular phylogenetic study (Wray et al. 1996), although their conclusions were challenged by Ayala et al. (1998), who argued that the divergence of metazoan phyla occurred no earlier than 670 million years based on their interpretation of molecular clocks. However, even if they are right, their analysis apparently does not rule out an earlier primitive metazoan emergence prior to the divergence of phyla.

A Geophysiological Model for the Evolution of the Biosphere

A geophysiological model for the evolution of the biosphere, summarizing the above discussion, is shown in figure 8-8.

The evolution of the biosphere is geophysiological in the sense that self-regulation arises from the coupling of biota and its inorganic surface environment. As emphasized in the discussion of the weathering process, the soil geomembrane is a critical site of this coupling. Markos (1995) pointed out

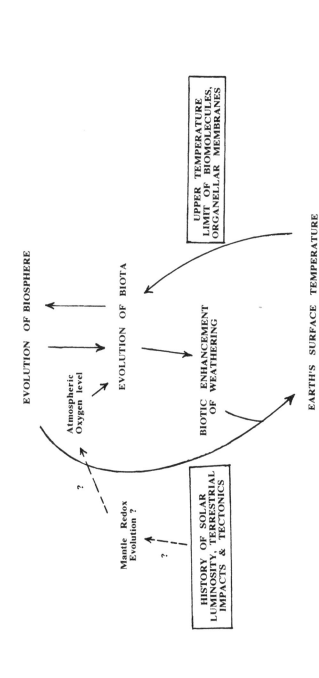

FIGURE 8-8.

Geophysiological model of evolution of biosphere. Constraints on feedback are in boxes. A mantle buffering of atmospheric pCO$_2$ was suggested by Kasting, Eggler, and Raeburn (1993).

that an extracellular matrix composed of mineral particles, cell exudates, organic debris, and biofilms is a second structure of biological communities. As such, it constitutes a mediating environment between the inorganic and organic bodies of the biosphere; the extracellular matrix of the soil is just such a mediator.

Surface temperature is one parameter that is regulated by the biosphere because the soil is the site where chemical weathering of CaMg silicates takes place. As previously argued from consideration of the carbonate-silicate biogeochemical cycle, progressive cooling arises from the circularity of the feedbacks, where biotic enhancement of weathering intensifies from both the direct effects of biotic evolution and regional/global effects (e.g., frost wedging in mountains), the latter also being a product of the evolving biosphere as a whole. There is of course an ultimate future limit to the self-regulating biosphere because solar luminosity steadily increases.

As shown in figure 8-8, the main abiotic constraints on the feedback of biospheric evolution to surface temperature are solar history and the Earth's tectonic and impact histories. The influence of solar history, particularly luminosity change as a function of time, has already been discussed (see chapter 3). The main factors of tectonics relevant to the first-order surface temperature history are the variation over time of volcanic/metamorphic outgassing of carbon dioxide (V) and land area (A). As introduced in simple models in chapter 3, higher V and lower A values relative to today's results in higher steady-state levels of carbon dioxide and temperature, all other factors being the same. In the past, all other factors were not the same, with lower luminosity tending to lower temperature, and lower biotic enhancement of weathering tending to raise temperature. The actual surface temperature at any time is determined by the combined effects of all factors contributing to achieving a steady-state level of atmospheric carbon dioxide, from the carbonate-silicate biogeochemical cycle (see discussion of climate models in the Appendix).

Have impacts had a major influence on the overall pattern of biotic evolution, and therefore temperature history, since the probable last opportunity for a global sterilizing event? By "overall pattern," I refer to the likelihood and timing of the emergence of the major groups of procaryotes, cyanobacteria in particular, and Eukarya and its kingdoms. Although there is growing evidence for impacts having produced mass extinctions in the Phanerozoic (Rampino et al. 1997), the Cretaceous-Tertiary (K-T) of course being the best known event, it is uncertain whether the overall pattern of biotic evolu-

BIOSPHERIC EVOLUTION

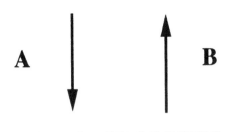

BIOTIC EVOLUTION

FIGURE 8-9.
Feedbacks between biotic and biospheric evolution.

tion has been dramatically affected by the Earth's impact history since 3.8 Ga. This possibility deserves closer study, particularly since several large impacts are likely to have occurred in the Precambrian (McKinnon 1997; Oberbeck et al. 1993).

We now focus in on the feedbacks between biospheric and biotic evolution (figure 8-9): feedback A entails the decrease in surface temperature, constraining at least microbial evolution, and the process of endosymbiogenesis, the history of emergence of cell types (e.g., Eukarya) and new kingdoms (i.e., the Protoctista, Fungi, Animalia or Metazoa, Plantae; see figure 8-10), because Eukarya and each of its new kingdoms have a decreasing upper temperature limit of growth corresponding to their approximate temporal order of emergence. Endosymbiogenesis begins with parasitic relationships between procaryotes, progresses to symbiotic relationships, and culminates in the intimacy of new cells and organisms composed of the endosymbionts (Margulis 1993, 1996; Margulis and Fester 1991; McMenamin and McMenamin 1994).

As previously argued, the history of land symbioses (table 8-2) may well be the key to the history of chemical weathering. Along with changes in productivity and oxygen-producing capacity, these changes in biotic evolution feed back into biospheric evolution (*B* in figure 8-9). A good case can be made for a progressive increase in biomass on the planet's surface (McMenamin and McMenamin 1994), particularly with the emergence of higher-plant forests (note the extensive coal deposits of the Phanerozoic age as well as forest soil of the Devonian; see Retallack 1997).

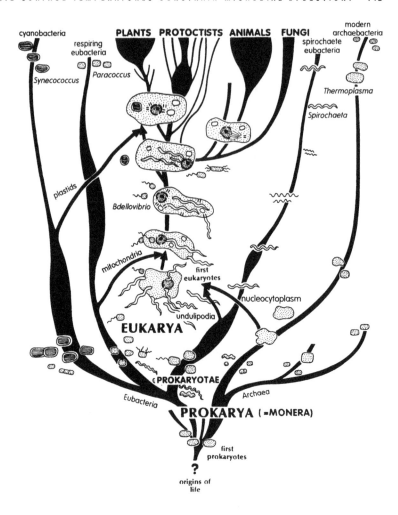

FIGURE 8-10.
The evolution of the five kingdoms, illustrating the role of endosymbiogenesis. Courtesy of Lynn Margulis.

From "Hothouse" to "Icehouse": The Increase in Diversity of Habitats for Life

With the long-term decrease of surface temperature over the past 3.8 billion years, a progressive expansion of the diversity of habitats has occurred (i.e., first hyperthermophiles, then thermophiles and mesophiles, and finally psychophiles, the organisms living near 0°C). Since the first appearance of ice on high mountains, perhaps even in the Archean, the diversification of habitats opened up new ecologic niches, with a cumulative retention of older habitats (e.g., there are still hot springs and the deep hot biosphere). As global temperatures decreased, particularly in the mid- to late Proterozoic, latitudinal differences in temperature and therefore a zonal differentiation of ecosystems ensued roughly parallel to the equator. The position and configuration of the continents also changed over time, affecting then as now the distribution of ecosystems.

As the atmosphere changed its composition in the mid-Proterozoic, becoming aerobic with the rise of oxygen, the area of surface aerobic environments expanded from the microenvironments (e.g., cyanobacterial mats) present before. Anaerobic environments persisted in soils, deep in the ocean, and in stratified bodies of fresh water, and finally in the gut of nearly emerged animals. If biotic productivity increased with the increase of macroeucaryotic algae in the oceans and on land, then the deposition of organic carbon in stratified water bodies could have perpetuated anaerobic conditions.

New pCO_2 (high, then low) and total pressure (first ≥ 10 bars, then low) environments progressively emerged since the origin of the biosphere at no later than 3.8 Ga. Again, the diversity increases as new environments are added, while old conditions persist (e.g., high pCO_2 and total pressure in the deep hot biosphere). In the case of pH, both very acid and basic and intermediate habitats were likely present even before the origin of life at hydrothermal vents. In Russell et al.'s (1994) interesting scenario, hot alkaline springs, generated in the early ultramafic crust by interaction with circulating seawater, react with acidic seawater at the sea floor to produce iron sulfide membranes. The latter are postulated to be the sites for catalyzed reactions producing protocells. A primordial pH diversity is consistent with the present wide range of optimal pH for growth of thermophilic procaryotes (Kristjansson and Stetter 1992). The coupling of the spectrum of pH habitats with lower temperatures and higher oxygen levels occurred with long-term cooling and the rise of atmospheric oxygen.

As the surface environments of the Earth changed over 4 billion years from a "hothouse" to an "icehouse," with a concomitant increase in diversity of habitats, the evolution of life naturally quickly filled these habitats with novel varieties. This scenario is consistent with evolutionary biological thinking, with the addition of a progressive increase of habitats globally over geologic time. Evolutionary biology up to now has only considered the progressive change of atmospheric oxygen in geologic time, as well as the opening up of new habitats within newly emerging organisms and ecosystems (e.g., within the digestive tracts of Metazoa and the forest canopy), not the possibility of progressive widening of the temperature, pCO_2, and total pressure regime of the biosphere.

The evolution of the biosphere does not optimize conditions for existing biota, unlike the original Gaia hypothesis. At least two catastrophes for existing life occurred: the well-known oxygen catastrophe and an earlier temperature catastrophe for thermophilic bacteria, once the likely colonizers of the ocean and land, now restricted to living in their thermophilic mode to hot springs, hydrothermal vents on the ocean floor, porosity in the first few kilometers of the crust, and hot water heaters.

Conclusions

The upper temperature limit for growth of organismal groups is apparently determined by the thermolability of biomolecules, organellar membranes, and enzyme systems. Aerobic microenvironments likely predated the rise of atmospheric oxygen (Archean oxygenic photosynthesis commencing at 3.5 Ga or earlier). In addition, atmospheric levels of oxygen sufficient for Metazoan metabolism may have preceded its emergence by at least 0.4 billion years. Thus, high Archean and early Proterozoic surface temperatures (50 to 70°C), inferred from the oxygen isotopic record in cherts and carbonates, could have held back the emergence of eucaryotes and Metazoa, with their upper temperature limits for growth corresponding to the ambient surface temperature at the time of emergence. A higher temperature limit for ancestral amitochondrial than mitochondrial eucaryotes would be consistent with suggestions that mitochondria are more thermolabile than nuclei.

The progressive increase in the biotic enhancement of weathering as a product of biotic and biospheric evolution intensified the carbon sink with respect to the atmosphere/ocean system, leading to the transition of climate from a hothouse in the early Precambrian to an icehouse in the Phanerozoic. By the late Proterozoic, the rise of atmospheric oxygen may have resulted in a substantial increase in terrestrial biotic productivity, which along with the onset of frost wedging substantially increased the biotic enhancement of weathering. Climate and life co-evolved as a tightly coupled system, constrained by abiotic factors (varying solar luminosity and the crust's tectonic and impact history). Self-regulation of this coupled system is a property of geophysiology.

Appendix: Climate Model Description

SURFACE TEMPERATURES ON EARTH SINCE THE EARLY ARCHEAN

These calculations, and those that follow in the next section, assume a balance between a land weathering sink for carbon dioxide and a metamorphic and juvenile outgassing source. The organic carbon burial sink is ignored. Organic carbon burial is irrelevant to abiotic models. As a first approximation, surface temperatures on Earth through the Archean are computed, ignoring possible variations in net organic carbon burial in the past.

Another carbon sink has been proposed—namely, the reaction of carbon dioxide with oceanic basaltic crust (see discussion in chapter 2), particularly in the Archean (Walker 1983, 1985; Staudigel et al. 1989; Veizer et al. 1989a, 1989b). Neodymium isotopes in Archean and Proterozoic sediments apparently support the notion that Archean ocean chemistry was dominated by the reaction with basalt, presumably at the ridges (Jacobsen and Pimentel-Klose 1988a, 1988b; for a contrary view see Alibert and McCulloch 1990). However, a thermal barrier to subduction of carbonate sediment or reaction product with basaltic crust (Des Marais 1985) may simply contribute an equivalent volcanic carbon outgassing source to the postulated sink, without affecting geochemical modeling of a land-weathering sink balancing the total assumed volcanic/metamorphic outgassing source, provided the variation of the latter parallels the fraction balancing the land-weathering sink (Berner 1990b).

First recall the equation given in chapter 6:

$$(1) \quad B = (P_{ab}/P_o)^\alpha \, e^{(\beta \Delta T)} \, e^{(\gamma \Delta T)} \, (A/A_o)(V_o/V)$$

An equation for the calculation of past surface temperatures (global mean) is derived for different assumed values of the ratio of biotic enhancements to that of the past (BR). In general,

$$(2) \quad B_t = (P_{ab}/P_t)^\alpha \, e^{\beta(\Delta T)} \, e^{\gamma(\Delta T)}$$

where B_t is the biotic enhancement at time t; (P_{ab}, T_{ab}) and (P_t, T_t) are the abiotic atmospheric pCO_2 and temperature and biotically enhanced values, respectively; $\Delta T = T_{ab} - T_t$; and the reference state is at time t and not the present as in equation (1). As before, T_o is taken as 288°K, and assumed values are $\beta = 0.056$, $\gamma = 0.017$, and $e^{(\gamma T)} \leq 2$.

Then,

$$(3) \quad B/B_t \; = \; \frac{(A/A_o)(V_o/V)\,(P_{ab}/P_o)^\alpha \; e^{\beta(\,T_{ab}-T_o)} \; e^{\gamma(\,T_{ab}-T_o)}}{(P_{ab}/P_t)^\alpha \; e^{\beta(\,T_{ab}-T_t)} \; e^{\gamma(\,T_{ab}-T_t)}}$$

where the numerator is simply from equation (1). Reducing to

$$(4) \quad BR \; = \; (A/A_o)(V_o/V)\,(P_t/P_o)^\alpha \; e^{\beta(\,T_t-T_o)} \; e^{\gamma(\,T_t-T_o)}$$

where $BR = B/B_t$.

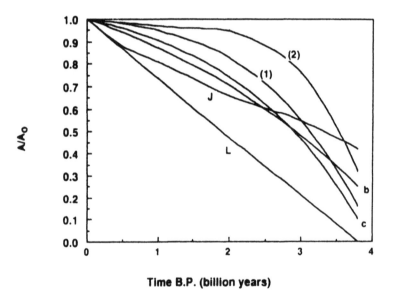

Time B.P. (billion years)

FIGURE 8-11.
Growth of continental land area (mass) versus time; A/A_o is the ratio of land area at time t to present. Curves b and c correspond to our models b and c, respectively, described in text. Curves (1) and (2) are from limiting tectonic/geochemical models of mean age of sediments and continents of Allegre and Jaupart (1985); curve J is from Jacobsen's (1988) inversion of Sm-Nd mass balance for the depleted mantle-continental crust system. Curve L, the linear growth of continental area through time, with $A = 0$ at 3.8 Ga, is taken as a lower limit to A/A_o as a function of age. Substantial early growth of the continents has received recent support from Bowring and Haush (1995), but a vigorous controversy is now in progress on this issue.

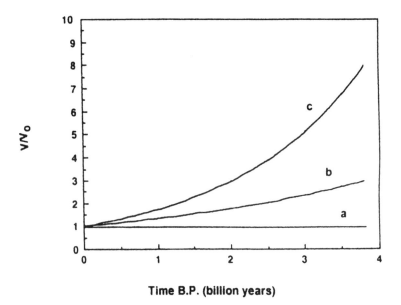

Time B.P. (billion years)

FIGURE 8-12.
Carbon outgassing rate relative to present rate (V/V_o) versus time (B.P.) for models a, b, and c.

A_o, A and V_o, V are defined here as the present and past continental land areas and volcanic/metamorphic outgassing rates (of carbon dioxide), respectively (figures 8-11 and 8-12). The following functions for volcanic outgassing, V, and land area, A, variation are assumed:

$$(5) \quad V = V_o\, e^{(\omega t)}$$

V parallels the decrease in radioactive heat generation in the Earth to present, where t is time (BP) in billions of years. We couple the rates of continental land area growth and volcanic outgassing:

$$(6) \quad (dA/dt)_t = -c(dV/dt)_t.$$

The rationale for assuming linkage of land area and outgassing rate is that both outgassing of carbon dioxide and continental crust generation are presumably a function of subduction and juvenile outgassing rates. Furthermore, intracontinental growth via underplating results in both outgassing and land area increases through uplift. This parameterization also simplifies

TABLE 8-3.
Parameters for Assumed Variation of Land Area (A) and Carbon Dioxide Outgassing
Rate (V) as a Function of Age (t)

Model	$(V/V_o)(A_o/A)$ at $t = 3.8$ Ga	cV_o	ω
a (constant)	$1 \times 1 = 1$	–	0
b (preferred)	$3 \times 4 = 12$	0.375	0.289
c (upper limit)	$8 \times 10 = 80$	0.129	0.547

model calculations. In any case, an alternative parameterization of land area
as a function of time (Allegre and Jaupart 1985; Jacobsen 1988; see figure
8-11) consistent with the isotopic evolution of the crust/mantle system, gives
similar model results for computed surface temperatures as a function of time
(see next section). The variation of land area and outgassing rate as a function
of time follows from models a, b, and c in table 8-3, with model a representing a constant V and A, model b the preferred intermediate variation, and
model c an upper limit to V and A since 3.8 Ga.

The relationship between surface temperature T_t and P_t, pCO_2 at any time
t, was computed from two different greenhouse functions:

(a) For $P_t < 0.03$ bar, an updated version of Kasting and Ackerman's
(1986) function, given in equation form by Caldeira (personal communication), with $Tt = f(P_t, S_t)$, where S_t is the relative solar flux at time t:

$$
\begin{aligned}
(7) \quad T_t = {} & 138.114 - 73.179(p) - 73.960(p^2) + 56.048(p^3) \\
& + 405.836(S_t) + 595.774(p)(S_t) + 385.004(p^2)(S_t) \\
& - 296.420(p^3)(S_t) - 316.907(S_t^2) - 839.205(p)(S_t^2) \\
& - 548.963(p^2)(S_t^2) + 461.125(p^3)(S_t^2) + 129.545(S_t^3) \\
& + 345.867(p)(S_t^3) + 251.4629(p^2)(S_t^3) \\
& - 261.438(p^3)(S_t^3)
\end{aligned}
$$

where $p = \log_{10}(P_t/1 \text{ bar})$

The maximum error is 1.95 K, the root mean square error is 0.65 K.

(8) $S_t = (1 + 0.0835t)^{-1}D^{-2}$, where p is the distance of the
Earth to the Sun (AU) [S_t as an $f(t)$ from Caldeira and Kasting
1992b]. (D values different from 1 AU will be pertinent to
modeling in chapter 11).

(b) For $0.03 > P_t \geq 0.0003$ bar, a slightly modified function from Walker et al. (1981) ($T_o = 288°K$ rather than 285°K, present $T_e = 255°K$ rather than 253°K):

$$(9) \quad T = 2T_e + 4.6(P_t/P_o)^{0.364} - 226.4,$$

where T_e is the effective radiating temperature of the Earth (°K; no greenhouse effect) at time t. The following relation for T_e from Kasting (1987) is used:

$$(10) \quad T_e = 255/(1 + 0.087t)^{0.25}$$

At low global mean temperatures (T_m), the use of T as above rather than the actual latitudinal distribution of temperature will exaggerate the feedback between climate and weathering rate because a much smaller contribution now comes from higher latitudes (White and Blum 1995), and latitudinal differences in temperature likely decrease with higher T_m. For model climates at higher T_m atmospheric pCO_2 levels than present values, the choice of T_m instead of the actual latitudinal distribution of temperature and rainfall probably becomes more realistic because latitudinal differences decline as T_m increases and may actually disappear at T_m equal to about 30°C (Hoffert, personal communication). Using the present continental latitudinal values of temperature and runoff (or precipitation) gives a much lower computed weathering rate (about 7%) compared with that derived from the T_m model, with the latitudinal distribution of runoff responsible for this difference (computed using temperature and runoff from Francois and Walker 1992, land area for 5° latitudinal strips from Sverdrup et al. 1942 with data from Kossinna 1921). However, the computed rate for actual latitudinal distribution of land, temperature, and runoff may well underestimate the actual silicate weathering sink because it does not take into account the silicate denudation rates by latitude. Another approach to this problem is as follows.

Assume using the present T_m instead of the actual latitudinal temperature variation and runoff pattern to get the present global weathering rate (Wn). Substituting T_m in the previously derived equation:

$$BR = (P_t/P_o)^{\alpha} e^{\beta(T_t - T_m)} e^{\gamma(T_t - T_m)}$$

TABLE 8-4.
Computed Model Results

T (°C)	Atm. pCO_2 (bars)	Time B.P (b.y.)	Required B_R a	b	c
70	2.56	3.5	658	82	25
60	1.22	2.6	301	82	43
50	.31	1.5	103	53	38

Because the computed Wn is greater than the actual global weathering rate, a lower P_o should be used to compensate, i.e., to bring the computed W_n down to the actual weathering rate. Using this hypothetical P_o will give higher computed B_R estimates.

SENSITIVITY OF SCENARIO H TO MODEL PARAMETERS

A value of $\alpha = 0.3$ is assumed, corresponding to abiotic weathering of CaMg silicates in an open system (Schwartzman and Volk 1989; Berner 1992, 1993); in a closed or nearly closed system, α most likely is > 0.3 (table 8-4). The choice of α, β (probable minimum★), and greenhouse function (pCO_2 is the probable minimum because cloud feedbacks are assumed absent) all make the computed B_R values lower limits. The normalization to the present global mean temperature rather than to the actual present latitudinal distribution of temperature and runoff also probably makes the computed B_R values lower limits (see previous discussion on the normalization problem). Probable higher impact-derived regolith, volcanic contribution to weathering, and the postulated sea floor weathering carbon sink (Francois and Walker 1992) relative to the present have the same effect on on the limit to B_R. Two possible factors would make the computed B_R values too high:

(a) reducing gas greenhouses in the Archean [although the $(Pt/Po)^\alpha$ term

---

★Regarding the $e^{\beta(\Delta T)}$ term, the more precise formulation using $[(1/288) - (1/T)]$ instead of a ΔT term, as well as the low activation energy (E) assumed (12.5 kcal/mole), give somewhat lower $e^{\beta(\Delta T)}$ values than for Brady and Carrol's (1994) lower limit on E (11.5 kcal/mole). Estimates of E are > 11.5 kcal/mole for CaMg silicates for most experiments near pH $= 7$ (Brantley and Chen 1995). Furthermore, E tends to increase as pH drops from the neutral range (Casey and Sposito 1992), again making $e^{\beta(\Delta T)}$ values lower limits for the assumed β ($= 0.056$) for high pCO_2.

would drop, only modest increases in the assumed α and β values would give the same or greater computed B_R values; see also previous discussion].

(b) greater net organic carbon burial now relative to the past (Des Marais et al. 1992). The present ratio is about 5.4 (data from Berner 1994) if V corresponds to volcanic/metamorphic flux only (weathering of kerogen is then substracted from organic C burial; see discussion in chapter 1). Let $R =$ ratio of silicate weathering to net organic C sink, assume $R = 5.4$ now. If all the carbon sink with respect to the atmosphere/ocean was silicate weathering at 3.5 Ga, then $(V_o/V_{3.5\,Ga})_{silicate\;sink} = 0.84\,(V_o/V_{3.5\,Ga})_{Total}$, since the (silicate weathering C sink/Total C sink)$_{now}$ = 0.84. Hence, under the above assumptions, the maximum reduction of B_R at 3.5 Ga is only 16% lower than the value computed without consideration of the influence of the organic C sink.

The effects of assumed model ratios of V_o/V and A/A_o being too low, as a result of the neglect of degassing from subduction of Cenozoic deposits of pelagic $CaCO_3$ (Volk 1989a; Caldeira 1991) and early continental crust formation (Bowring and Housh 1995) results in the computed B_R values again being minimal. On the other hand, pelagic $CaCO_3$ deposition also may have occurred in the Archean (Grotzinger 1994). Subduction of exogenic carbon into the mantle and its reappearance in volcanic outgassing (Hauri et al. 1993) has probably decreased the variation of V_o/V ratios over time. From the observed carbon and helium isotopic compositions and $CO_2/^3He$ ratios of volcanic gases, Sano and Williams (1996) estimated that 60% of the present degassed carbon flux comes from subducted sediment carbon (mainly carbonate).

A rough material balance of recent pelagic carbonate deposition and subduction/decarbonation fluxes indicates that about one third of pelagic carbonate is now degassed, implying carbon loss to the mantle (Caldeira 1991). This conclusion is supported by a recent study (Nishio et al. 1998). Furthermore, it is possible that no exogenic carbonate reservoir existed 3.8 billion years ago as a result of low pH oceans and rainwater. A shift from the contribution to outgassing of juvenile to metamorphic carbon (in subduction zones) from the Precambrian to now would tend to stabilize the flux as a function of time. Clearly, a greater understanding of the carbon geodynamic cycle is needed before results from modeling the carbonate-silicate cycle back into the Precambrian can be accepted as definitive. Nevertheless, useful limits can be obtained with modeling.

The Carbonate-Silicate Geochemical Cycle at 3.8 Ga, for an Abiotic Earth Surface Just at the Transition to Biotic Colonization

I assume here that the late bombardment, with its potential for continual large sterilizing surface impacts, was over by 3.8 Ga, thus taking this time as the transition time to biotic colonization of the early continents (Oberbeck and Mancinelli 1994).

With the assumptions made above, the product of outgassing rate and land area ratios, $(V/V_o)(A_o/A)$, at $t = 3.8$ Ga for two limiting models (models a and c) and a preferred model (b) is given in table 8-3. The variation of A/A_o, V/V_o, and $(V/V_o)(A_o/A)$ for the above models is shown in figures 8-11, 8-12, and 8-13, respectively. The upper limit of the outgassing ratio, V/V_o, is taken as 8 for the early Archean, approximately 30% higher than an estimate for 3 billion years ago made from a consideration of heat generation rates and depth of eutectic melting and source volatile outgassing (Des Marais 1985). The lower limit to land area is taken as 10% of the present value

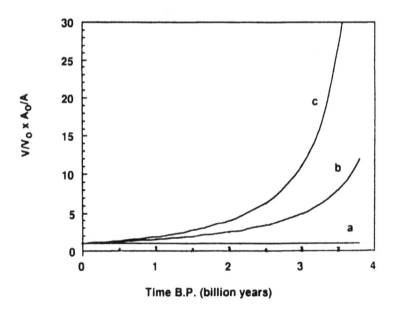

FIGURE 8-13.
$(V/V_o) \times (A_o/A)$ versus time for models a, b, and c. Value for model c at 3.8 Ga is 80 (not shown).

(Allegre and Jaupart 1985 gave values of 16% to 32% of the present continental mass at this time), and an upper limit to land area as 100% of the present value.

Isotopic dating reveals continental crust in existence at 4 to 4.3 Ga (Bowring et al. 1989). Armstrong (1968, 1981, 1991) has argued for a constant volume continental crust in existence for at least 4 billion years, a model that apparently fits the isotopic data (Nd, Sr, Pb, etc.) as well as the net continental growth models. In Armstrong's model, the rate of continental growth equals its rate of destruction in subduction zones. This controversy was revived by Bowring and Housh (1995), who favor the theory of a substantial early continental crust. Conversely, Taylor and McLennan (1995) and Mc-Culloch and Bennett (1994) favor the notion of more gradual growth, with less than 10% of present volume at 3.8 Ga (60% by 2.5 Ga). Vervoort et al. (1996) have suggested that neodymium isotopes in the most ancient Archean gneisses—the basis of Bowring and Housh's (1995) argument—were probably reset by metamorphism, based on Hf isotopic pattern in zircons. From a consideration of the Nb/U ratio of Archean basalts compared with the primitive and present day mantle ratios and other evidence, new support has emerged for rapid continental growth early in Earth history, with most of the present mass in place by 2.7 Ga (Sylvester et al. 1997; Hofmann 1997; Sylvester 1998). However, taking into account the apparent variation of the U/Th ratio in the depleted mantle over geologic time, Collerson and Kamber (1999) favor a growth model similar to that of McCulloch and Bennett (1994). This interpretation is in variance with those favoring early continental crust, with little growth since 3.5 Ga, based on detailed geochemical studies of Archean crust (Calderwood 1998; Sylvester et al. 1998).

Note that if higher sea-floor spreading rates prevailed in the Archean, the sea level should have been higher, thus favoring a lower estimate for land area relative to continental mass than today (Drever et al. 1988).

Extreme limits for $(V/V_o)(A_o/A)$ are taken for the assumed values at 3.8 Ga. Model a, with the constant $A = V = 1$, is a lower limit to $(V/V_o)(A_o/A)$. Although a constant or slowly decreasing outgassing rate to present may seem implausible, it is compatible with estimates of the total present exogenic carbon inventory and carbon outgassing fluxes derived from measured $C/^3He$ in MORB glasses and ^3He outgassing fluxes (Marty and Jambon 1987; Marty 1989). A $(V/V_o)_{3.8\ Ga} = 3$ assumes a heat generation rate of four times the present rate, with komatiitic oceanic ridge formation dissipating

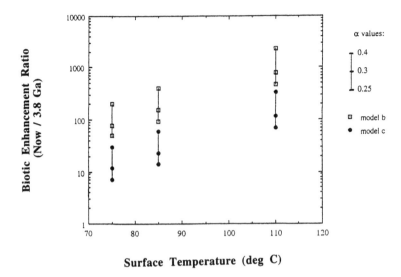

FIGURE 8-14.
3.8 Ga model results: biotic enhancement of weathering ratio, now to 3.8 Ga versus surface temperature (75, 85, and 110°C). Computed results shown for model b and c for
three α values.

30% more heat than the basaltic ridge at the same spreading rate (Nisbet
1987). To sum up, input parameters for our 3.8 Ga model are the ratios
of outgassing rates and land areas, $(V/V_o)_{3.8\ Ga}$ and $(A_o/A)_{3.8\ Ga}$, respectively,
atmospheric pCO_2 and hence temperature (from a modified version of Kasting and Ackerman's 1986 greenhouse function described in the section on
climate modeling), with the ratio of the present biotic enhancement factor
to that at 3.8 Ga (B_R) as the output. Three temperatures corresponding to
growth conditions for thermophiles/hyperthermophiles were chosen.

Calculations for the 3.8 Ga model give conditions values of B_R = 7 to
2250 (preferred model b for variation of outgassing and land area, 49 to 2250;
model a results, which give even higher B_R values, are not given here) for
temperatures of 75 to 110°C for an abiotic Earth surface just at the transition
to biotic colonization (figure 8-14). If the Earth's atmosphere at this time
contained 10 to 20 bars of carbon dioxide, surface temperatures could have
approached 100°C (Kasting 1989), near the approximate upper temperature
limit of hyperthermophilic microbes.

The range of B_R values computed from the 3.8 Ga model is consistent with limits on the present biotic enhancement of weathering (B), derived from experimental and field studies of weathering; in other words, values on the order of 100 to 1000 if the assumed estimate of $(V/V_o)(A_o/A)_{3.8\,Ga}$ is not much higher than the model b value (even model c is probably consistent with $B = 100$ to 1000 because the computed B_R values are likely lower limits for the reasons outlined in the section on the sensitivity of scenario H).

We previously outlined a scenario for the colonization of land by extreme thermophiles soon after the origin of life (Schwartzman and Volk 1989, 1990), leading to the accelerated removal of carbon dioxide via reaction with silicates and decrease in temperature. A clarification to this scenario, already discussed, is that the origin of life may have occurred as early as 4.2 Ga, but the first likely opportunity for the persistence of surface life was at about 3.8 Ga. Temperatures for growth of extant thermophiles/hyperthermophiles range from 74°C (upper limit for cyanobacteria) to 85°C (methanogens) to 105°C for *Pyrodictum* (upper limit 110°C) (Brock 1986). Chemical weathering intensities (fluxes/unit land area) relative to the present rate $\{(V/V_o)(A_o/A)\}$ for the 3.8 Ga model, assuming steady state, range from 1 to 80 (model b, 12). Are these unreasonably high? If the present global physical denudation rate is some six times the chemical rate (Holland 1978), the maximum chemical denudation rate possible at the present mean continental uplift rate is six times the present rate. Thus, at 3.8 Ga, minimum mean uplift rates of 0.17 to 13 (for model b, 2) times the present rate are needed for a steady-state carbon dioxide level in the atmosphere/ocean system. If the mean uplift rates in the early Archean were significantly higher than today's rate (E. Nisbet, personal communication), then higher chemical denudation rates would be possible. Alternatively, a steady state was not achieved until later in the Precambrian as volcanic emissions declined and land areas increased.

Another barrier to achieving steady state was the sheer amount of carbon dioxide in the atmosphere at 3.8 Ga (for 10 bars, most of the carbon dioxide would be in the atmosphere as a result of a low pH ocean; see Walker 1985). The removal time of a 10 bar carbon dioxide atmosphere is estimated to be on the order of 10^8 years computed by dividing an exogenic inventory corresponding to 10 bars (2×10^{21} moles) by $\{(V/V_o)_{3.8\,Ga} \times V_o\}$, taking $(V/V_o)_{3.8\,Ga} = 3$, V_o as the volcanic carbon flux (6×10^{12} moles/year; Berner et al. 1983) to the atmosphere/ocean. Because land area, volcanic outgassing rate, and solar luminosity change significantly over 10^8 years, a steady-state approximation is not entirely justified. However, the steady-state model is used for its computational simplicity because the focus is on the role of biotic enhancement of weathering. Thus, the model calculations represent attractors toward which the system is relaxing at any point in time, and not the actual values, which could lag behind by several hundred million years. A lag time of this magnitude would apply until the mid-Archean, when atmospheric carbon dioxide levels decreased dramatically, making the steady-state model a much better approximation to real behavior (e.g., see the Phanerozoic model of Berner 1990a).

Transient models have been computed by Tajika and Matsui (1990) and Walker (1991). Tajika and Matsui (1990, 1992, 1993) assume that initially all the Earth's carbon (aside from that in the core) was in the surface reservoir because of the partitioning of carbon between the proto-atmosphere and early magma ocean. Thus, in their model, the degassed carbon during Earth history comes from the release from subducted carbonate and a return flux from a regassed component derived from the subducted carbon lost to the mantle. However, because the speciation of carbon in the mantle is not well understood (SiC may be an important phase according to recent evidence; see Leung et al. 1990), it may be premature to conclude that the Earth's early mantle lacked levels of carbon high enough to supply a significant primordial flux through degassing over geologic time. Walker (1991) supports this viewpoint. Furthermore, the postulated regassing in the Archean is perhaps implausible in view of the apparent thermal barrier (Des Marais 1985; Abbott and Lyle 1984).

Theory of self-organization (Kauffman, Ortoleva). Lovelock's views on Gaia as a disequilibrium system; why geochemistry alone can generate disequilibrium of the crust and atmosphere. Geochemical versus geophysiological self-organization of the biosphere, a thermodynamic approach using the net entropic flux from the Earth's surface as an index. The life span of the biosphere; the future breakdown of self-organization (carbonate-silicate system goes to chemical equilibrium). Homeostasis redux. The self-organizing biosphere: theses on biospheric evolution.

Is the Earth's biosphere a self-organized system? How different is the Earth's surface system from an equilibrium state? Would the Earth's surface system be self-organized without the presence of life? How can the self-organization of the biosphere be measured and tracked since the origin of life? Would this characterization give any insight into the evolution of the biosphere and life? These are some of the questions raised in this chapter. I start with a discussion of the regulation of the Earth's surface system, the biosphere, that is, the space-time inhabited by life, consisting of the atmosphere, hydrosphere, soil, and upper layers of the crust (deeper than we have previously thought). In particular, I focus on a thermodynamic treatment of the surface system's evolution. Previous thermodynamic approaches to the biosphere's self-organization are discussed. Finally, a potential indicator of the biosphere's self-regulation is proposed, namely, the trend in the net entropic flow from its effective physical boundary (i.e., the boundary of the biota itself), the surface of the ocean and land, compared with the trend for an abiotic surface.

The origin of the Gaia concept is rooted in Lovelock's realization that the

Martian atmospheric composition should be indicative of the presence or absence of an indigenous biota:

> If the planet were lifeless, then it would be expected to have an atmosphere determined by physics and chemistry alone, and be close to the chemical equilibrium state. But if the planet bore life, organisms at the surface would be obliged to use the atmosphere as a source of raw materials and a depository for wastes. Such a use of the atmosphere would change its chemical composition. It would depart from equilibrium in a way that would show the presence of life. (Lovelock, 1990)

Smolin (1997) repeated this argument. However, this view ignores the possibility of purely geochemical nonequilibrium condition of a planetary surface system resulting in steady-state properties, such as atmospheric carbon dioxide level, that are different from chemical equilibrium values. [It is interesting that Hutchinson (1954), considering the Urey reaction, concluded that several factors on the Earth operate to preclude the likely attainment of an equilibrium level of atmospheric pCO_2]. Walker et al.'s (1981) geochemical climatic stabilizer is a model of the operation of just this kind of global geochemical self-organization.

Smolin (1997) characterized the evolution of the biosphere as "the self-organization of a complex system with many interacting components" (p. 149). He defined a self-organized, nonequilibrium system as "a distinguishable collection of matter, with recognizable boundaries, which has a flow of energy, and possibly matter, passing through it, while maintaining, for time scales long compared to the dynamical time scales of its internal processes, a stable-configuration far from thermodynamic equilibrium. This configuration is maintained by the action of cycles involving the transport of matter and energy within the system and between the system and its exterior" (p. 155).

Here I define self-organization as the autonomous passage of a system from an unpatterned to a patterned state without the intervention of an external template. Furthermore, "self- organization requires feedback and disequilibrium, both of which occur in geochemical systems" (Ortoleva et al. 1987). Examples of geochemical self-organization are exceedingly common. They include oscillatory zoning in plagioclase crystals from igneous rocks, metamorphic differentiation (i.e., alternating layers of mineralogically distinct assemblages produced by metamorphism of homogeneous parent rock), banded agates (figure 9-1), and stylolites (Ortoleva 1994; Merino 1992).

Self-organization of the biosphere is indicated by the following characteristics:

1. The very existence of the living biota, in necessary disequilibrium with the nonliving part of the biosphere.
2. Steady states, not equilibrium levels, of carbon dioxide, oxygen, and so forth in biogeochemical cycles, a product of dynamic feedback relationships, both positive and negative. The steady-state level of atmospheric carbon dioxide is a direct outcome of the carbonate-silicate climatic stabilizer, whether it be geochemical or geophysiological. The computed equilibrium levels of carbon dioxide and corresponding equilibrium temperatures for all past geologic time are dramatically different from the steady-state levels, as will be shown in what follows.
3. The differentiation of the upper crust into its soil interface with the atmosphere—itself differentiated into topsoil, subsoil, and so forth (Nahon 1991)—and its underlying biogenic sedimentary layers.

Thus, the Earth's surface system can be self-organized, whether it be geochemical or geophysiological, because it possesses feedback and disequilibrium (Ortoleva et al. 1987). If the early pressure-cooker atmosphere at 3.8 Ga was in steady state, geochemically balanced between a volcanic source and weathering sink, assuming plausible limits on continental area and volcanic outgassing rates, then the present biotic enhancement of weathering (B) is likely two orders of magnitude (see discussion in chapter 8). This requirement for a high present B follows directly from the need, under steady-state abiotic conditions, to balance the higher volcanic flux of carbon by weathering on somewhat smaller continents, some 4 billion years ago, necessitating high atmospheric carbon dioxide levels (and correspondingly high surface temperatures) to speed up weathering, work now done by biotic enhancement at a much lower carbon dioxide level.

A much warmer Earth surface in the Archean/early Proterozoic, particularly in the early Proterozoic without the plausibility of a reduced gas greenhouse, supports a high present biotic enhancement of weathering factor as a result of higher plant colonization of the continents (see chapter 8).

FIGURE 9-1.
Photos of banded agate (top) and banded rhodochrosite (bottom) I took at the Smithsonian Museum of Natural History, DC.

Entropic Flow from the Earth's Surface

If the biosphere is a self-organized system, can its dynamics and evolution be modeled by an understanding of its entropic relationships with its environment? Ebeling (1985) and Lesins (1991), among others, have pointed out that the net entropic flow from the Earth's surface and atmosphere is a necessary condition for biospheric self-organization. The flow is possible because of the existence of a heat sink in outer space; low entropy solar radiation comes into the biosphere, and high entropy heat radiation leaves it to the heat sink.

I suggest here that the cooling of the Earth's surface since the origin of life is a result of the progressive self-organization of the biosphere, to be quantified by the net entropic flow from the land and ocean's physical surface, the upper surface of the biosphere with the actual presence of the biota, not including the atmosphere with its incidental living occupants, namely, birds, bats, flying humans, microbial spores, and so forth, a negligible part of the biota's mass. The interaction of solar radiation with the Earth's surface, or any surface that is not perfectly reflecting (i.e., with albedo $= 1$) is an irreversible process, hence $dS/dt > 0$, where S is entropy and t is time. Incoming solar radiation is characterized by low entropy (high temperature) and the net outgoing radiation by high entropy (low temperature). The temperature is related to the frequency of radiation, which is proportional to photon energy. Because the same amount of energy contains fewer photons in the form of solar radiation than for terrestrial radiation, solar radiation is more organized (higher temperature) than terrestrial radiation (see discussion in Peixoto et al. 1991). This change in radiation entropy is a result of its absorption and reradiation by matter because the identity of any one photon is lost upon absorption.

Lesins (1991) attempted to see a Gaian signal in the outgoing radiation stream from the top of the atmosphere using satellite data. He concluded that such a signal was not apparent because the meteorologic and nonliving conversions at the Earth's surface and in the atmosphere dominate the entropy signal. However, he suggested that the "correct thermodynamic test of Gaia is to compare the outgoing entropy fluxes between the current Earth and one devoid of life" because life has presumably created a climate much different (e.g., chemical composition) than one created abiotically. Ulanowicz and Hannon (1987) also have proposed searching for the entropic signature of life using remote sensing data, but from the surface, not from the top of the atmosphere.

I suggest here that the entropic flow from the Earth's surface should be a much more sensitive index of the biosphere's self-organization than the flow at the atmosphere/space interface, which is very close to that expected for the Earth as a black body radiating to space.

The energy budget of the Earth-atmosphere system is shown in figure 9-2. The corresponding entropic budget is found in the work of Peixoto et al. (1991; see also Aoki 1988).

We now compute the net entropy flow across a boundary just above the Earth's surface. The flux is a sum of separable fluxes, taken from the Earth's energy balance (see figure 10–1). The net entropy flow, $(dS_{surface}/dt)$, from the surface is therefore:

$$(1) \quad (dS_{surface}/dt) = -(E_{sun}/T_{sun}) - (E_{atmIR}/T_{atm}) + (E_{sensible}/T_{surface}) + (E_{latent}/T_{surface}) + (E_{surfaceIR}/T_{surface})$$

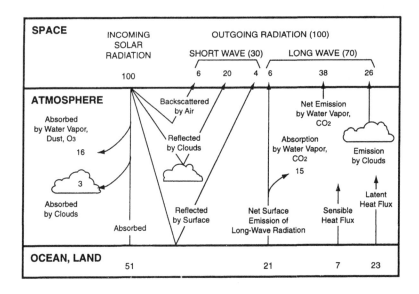

FIGURE 9-2.
Radiation budget in the Earth-atmosphere system. Courtesy of National Academy of Sciences, 1983.

where the E's are the various energy fluxes in positive values of watts m^{-2}, and T's are the temperatures of the effective surface producing the fluxes. The total nonsolar energy flux crossing the surface must by energy balance equal the flux received from the sun, E_{sun}:

$$(2) \quad E_{sun} = -E_{atmIR} + E_{sensible} + E_{latent} + E_{surface\,IR}$$

Furthermore, any reasonable assumption for $T_{atm} < T_{surface}$ makes a negligible difference in the entropy flux associated with this total nonsolar energy flux. For example, if T_{atm} is taken as 10°C cooler than $T_{surface}$ (T_{atm} being an effective radiating temperature from atmosphere to surface, which is actually a sum of radiations from different, primarily near-surface altitudes), the effect on the total nonsolar entropy flux is only several percent. Using the energy balance, and $T_{atm} \approx T_{surface}$, equation (1) becomes

$$(3) \quad (dS_{surface}/dt) = -(E_{sun}/T_{sun}) + (E_{sun}/T_{surface})$$

Because $T_{sun} = 5760°K$ and $T_{surface} = 288°K$, for the purposes of this calculation we can drop the solar contribution to the entropy flux and finally write

$$(4) \quad (dS_{surface}/dt) = (E_{sun}/T_{surface})$$

We will use equation (4) for the calculations in this chapter. E_{sun} is proportional to solar luminosity (L) at any time t. L has increased 25% in the past 4 billion years. We make the following assumptions in our model calculations:

1. The Earth surface has maintained a constant albedo (Kasting and Toon 1989).
2. Radiative and local thermal equilibrium obtains at the surface (Peixoto and Oort 1992). As discussed above, the small correction to the net entropic flux arising from reradiated atmospheric infrared with $T < 288°K$ is neglected.
3. A standard model of L variation (see discussion in chapter 2).
4. The nonclassical part of radiative entropy (Lesins 1990; Essex 1984) can be neglected because it is small compared with the classical part.
5. The entropic production by photosynthesis accounts for only 0.1% of

the surface flux (Lesins 1991), whereas that from weathering and surface chemical reactions is only 10^{-8} of the net flux; thus, both contributions can be neglected.

I now proceed to compute the net entropic flow from the surface, dS/dt, as
a function of time, normalizing it to the flow corresponding to an Earth with
no greenhouse effect. This normalization procedure corrects for temporal
changes in entropic flow resulting from the variation of solar luminosity
alone over time. Because the normalization procedure cancels out E_{sun}, then
$(dS/dt)_{normalized} = (T_e/T_{surface})$. T_e is the effective temperature corresponding
to a no-greenhouse Earth. We proposed that biospheric self-organization
is revealed in the comparative trends of the normalized entropic ratio
corresponding to biotic, abiotic, and equilibrium temperature histories
(Schwartzman et al. 1994). These trends will be discussed in the next section.
Assumed model temperatures and computed results are shown in table 9-1,

TABLE 9-1.
Surface Temperatures (°K)

Age (Ga) Scenario	T_{biotic} A	B	$T_{abiotic}$[a]	$T_{equilibrium}$	T_e	$(dS/dt)ng/(dS/dt)ng_{now}$[b]
−2.5			372	371	271	1.20
−2			344	325	268	1.15
−1.5	[c]		333	304	264	1.11
−1	300	300	329	291	261	1.07
−0.5	291	291	327	283	258	1.03
0	288	288	326	276	255	1.00
1	293	298	329	266	250	0.94
3.5	313	346	350	251	239	0.82
3.8[d]	356	356	356	250	237	0.81

[a]Assumes $B = 100$, $\alpha = 0.3$ with modified Walker greenhouse function used for $3.8 > t$
> 0 (Schwartzman and Volk 1991a), Caldeira and Kasting (1992b) greenhouse function
for $t < 0$.
[b]ng, no greenhouse.
[c]The temperature increases rapidly after >1.0 Ga in Caldeira and Kasting's model (1992b)
as a result of holding pCO_2 atmospheric constant at 10^{-6} bars.
[d]Just prior to origin of life.

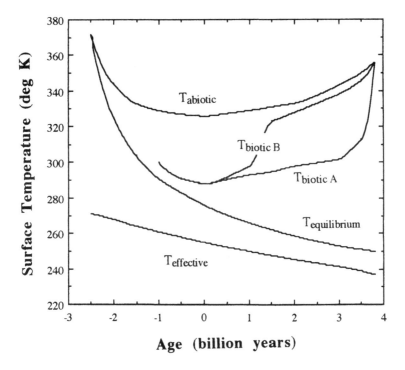

FIGURE 9-3.
Surface temperatures versus age (negative age corresponding to future time).

and figures 9-3 and 9-4. Details of the computation are provided in the Appendix.

The assumed temperatures for a biotic Earth surface are derived from geologic constraints, with two scenarios indicated. Scenario A corresponds to the conventional low-temperature history (Kasting and Toon 1989), whereas scenario B assumes a warm Archean/early Proterozoic as theorized in chapter 8. I assume future temperatures on a biotic Earth follow the path computed by Caldeira and Kasting (1992b) (see Appendix).

Model Results and Discussion

The main result of entropic flow calculations is the higher normalized entropic ratio from the origin of life to present corresponding to the biotic surface (scenarios A and B) compared with the abiotic (see figure 9-4), a

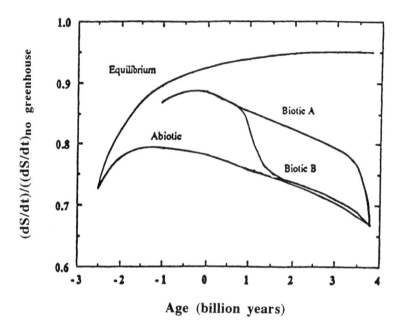

FIGURE 9-4.

Ratio of computed entropic flux from surface to entropic flux from a no-greenhouse Earth surface, $(dS/dt)/(dS/dt)_{\text{no greenhouse}}$, versus age (negative age corresponding to future time).

direct result of the lower temperatures for the biotic scenarios. The ratio corresponding to equilibrium temperatures for the Urey reaction shows a steady decline over the same time. A further insight into biospheric self-organization may come from comparing the ratio of net entropic flow for a biotic Earth surface now relative to the abiotic surface just prior to the origin of life at 3.8 Ga (assuming a surface temperature of 356°K). This ratio, $(dS/dt)_{\text{now}}/(dS/dt)_{3.8\,\text{Ga}}$, is 1.65, whereas without a greenhouse effect $(T_{\text{surface}} = T_e)$ the ratio is 1.24. A possible index of the cumulative self-organization of the present Earth surface system, normalizing the effect of luminosity change from 3.8 Ga to now, is the ratio of the former value to the latter, 1.33. It is important to emphasize that the index for self-organization of the surface system is not the net entropic flow itself, but its normalized trend over time because an inert Earth, with no atmosphere, would radiate at black body temperature with a higher net entropic flow than a biotic Earth surface. The black body net entropic flow simply expresses the irreversibility of surface radiative absorption and reradiation, while it has increased monotonically since 3.8 Ga (see Table 10–1; note trend in $(dS/dt)_t/(dS/dt)_{now}$ for the no-greenhouse case).

The biosphere has increased the net entropic flow by its reduction of atmospheric carbon dioxide and cooling of the Earth's surface, resulting from the biotic enhancement of weathering. In the geophysiological interpretation, the biota progressively speeds up weathering to present with the surface temperature approaching the equilibrium temperature while the biosphere self-organizes. Equilibrium conditions for the Urey reaction can be viewed as an attractor state for the biosphere. Self-organization in the sense previously defined is expressed by the achievement of lower steady-state atmospheric carbon dioxide levels (and therefore surface temperatures) obtaining under biotic conditions than under abiotic conditions, at the same weathering flux at a given time under biotic conditions as abiotic conditions. The key site for this self-organization is at the interface between land and atmosphere, the soil, the Earth's geomembrane (Arnold et al. 1990), where carbon is sequestered by its reaction (as carbonic and organic acids) with calcium magnesium silicates. The occurrence of differentiated soil goes back to the Archean (Retallack 1990) and is the critical material evidence for biospheric self-organization. This biotic invasion into the surface of the continental crust constitutes an ever-expanding front of high surface area/land area, with progressive colonization of land and evolutionary developments culminating in the rhizosphere of higher plants. Written in another context, Haldane's (1985) remarks on the "struggle to increase surface in proportion to volume" aptly describe this history of progressive self-organization.

Returning to the concept of equilibrium as an attractor state for the biosphere, does this mean that the Earth's surface is becoming more life-like as it approaches the attractor? Yates' (1987a) views are interesting in this connection: "Biological systems operate in the domain of irreversible near-equilibrium thermodynamics."

This near-equilibrium state is brought into being by the catalytic role of enzymes. Similarly, the biotic catalytic role in weathering the crust brings the surface temperature closer to the equilibrium temperature of the Urey reaction. Mae-Wan Ho (1993, 1995) has proposed a new theory of bioenergetics which emphasizes the role of energy storage in living systems that are dissipative isothermal engines (not Carnot cycles; see Morowitz 1978), with internal domains of local equilibrium.

Morowitz (1992) has pointed out that the equilibrium state is a powerful attractor in matter recycling, a consequence of nonequilibrium systems maintained by energy flows (low entropy energy in, high entropy heat out). Could not the approach over geologic time to the attractor equilibrium state of the biospheric surface as indicated by the level of atmospheric carbon dioxide, along with the expansion of energy storage in the biosphere with the rise of vascular plant ecosystems, be evidence of the biospheric surface system becoming more life-like? Note that the present reservoirs of carbon in soil and biomass are three to four times the mass in the atmosphere (see figures 2-1 and 2-2).

Is there a one-to-one relationship between the computed entropy change (a direct consequence of surface cooling) and the entropy change arising from the process of self-organization? In the sense that I have defined self-organization, it includes an informational aspect entailing the ordering of the biosphere, expressed, for example, by the differentiation of soil and layering of the crust, an ordering with a corresponding entropy production. Although I have not attempted to quantify the latter, I conclude that the total entropic production must in principle be greater than the radiative entropic component. How much greater is a question left for future consideration; as already pointed out, the direct entropic production from sequestering of atmospheric carbon by photosynthesis and weathering, as well as other surface chemical reactions, is a small fraction of the net radiative entropic flux. However, I argue that the inferred cooling of the Earth's surface is a direct consequence and index of the process of self-organization outlined above; in this sense the computed radiative entropic flux tracks the process of self-organization. It is a correlative, even perhaps an entirely coincidental measure, not necessarily a causal one, of this process.

For the case of a hypothetical Earth without surface cooling, biology would not have much influence on temperature; presumably thermophilic Archea and Eubacteria could inhabit such an Earth, but not low-temperature life. On this Earth, self-organization of the surface system would be purely geochemical, and presumably less developed than on our Earth.

On a future Earth both the abiotic and biotic temperatures converge on the equilibrium temperature by about 2.5 billion years from now (figure 9-3). The normalized entropic fluxes for the abiotic and biotic cases peak at 0.5 to 1 billion years in the future and then decrease (figure 9-4). This decrease corresponds to the breakdown of self-organization of the Earth's surface system, a result of the increase of solar luminosity overwhelming the regulatory capability of the biosphere, with the carbon dioxide greenhouse effect disappearing. At this point, temperature increases arise from a water greenhouse only with chemical equilibrium being achieved from the temperature effect alone. Lovelock and Whitfield (1982) noted just this eventual breakdown of the biosphere's capacity to regulate climate, because according to their estimate, the pCO_2 atmospheric pressure will reach the lower limit (150 ppm) tolerable for photosynthesis in some 100 million years. Caldeira and Kasting (1992b) have revised Lovelock and Whitfield's calculations, taking into account more complex greenhouse and biologically mediated weathering models and that C4 photosynthesis can persist to levels of atmospheric carbon dioxide below 10 ppm. However, some thermophiles might be expected to survive as temperatures rise to above the limit for eucaryotes (50 to 60°C). According to Caldeira and Kasting's calculations, within 1.5 billion years, photodissociation of water, with loss of hydrogen to space, would be a significant process, leading to the loss of the hydrosphere in another billion years.

Even if the biotic enhancement of weathering is now and has been in the past modest, an unlikely condition as previously argued, the Earth's surface system would still be self-organized, but purely geochemical; in this case a pressure-cooker atmosphere at 3.8 Ga could not have been in steady-state with early continents.

The progressive self-organization of the biosphere also may be indicated by rising atmospheric oxygen levels, in growing disequilibrium with the crust, since the Proterozoic (Swenson 1990; Swenson and Turvey 1991). This trend, however, has had no direct effect on the trend of entropic flow from the surface; its effect may have been indirect in its contribution to increased biotic enhancement of weathering under aerobic conditions (possible role of ozone shield, aerobic microbial communities in soil, emergence of higher plants).

It is interesting to compare the pattern of variation of the net normalized entropic flow out of the Earth's surface system since its birth to that of analogous functions (e.g., rate of entropic production/unit area) in the evolution of other thermodynamically open systems such as living organisms (Lamprecht and Zotin 1978; Aoki 1992). Both functions exhibit a growth stage, level out, and then decline with the breakdown and ultimate collapse of the self-organized system. However, it is important to note the distinctiveness of the Earth's surface system. The net entropic flow out is essentially from radiative interaction at its interface with the atmosphere, not directly from internal irreversible processes as in other thermodynamically open, self-organized systems, such as heat generation in, say, a mouse. Yet as argued above, the self-organization of the biosphere has been responsible for just this trend in entropic flow. Could this expression of biospheric evolution reflect a deeper commonality of development characteristic of self-organized thermodynamically open systems? Or alternatively, is the inferred pattern of the variation of net entropic flux a trivial expression of the unique temperature evolution of the Earth surface system, without any deeper significance?

Views on the applicability of thermodynamic concepts to organisms and biospheres range over a wide spectrum. Morowitz (1992) contends that entropy production cannot be validly used as an indicator of organizing tendency for these far-from-equilibrium systems because entropy can only be validly defined at equilibrium (but see the illuminating discussion of this issue by Bricmont 1996). In contrast, Swenson (1990) and Swenson and Turvey (1991) have argued that a principle of maximum entropy production predicts the very emergence of these self-organized systems because the export of entropy is a necessary condition for self-organization. Our computed trend in the normalized entropy flux is consistent with Swenson and Turvey's general prediction for the evolution of such systems, but this fact does not necessarily validate their theory.

Entropy flow out is a necessary but not sufficient condition for the emergence of self-organized systems. Sufficient conditions include those present initially, as well as at the boundary during the existence of the self-organized system. For example, if the origin of life is a deterministic process that occurred in hydrothermal systems near 100°C (Wachtershauser 1988), the sufficient conditions include those present in the hydrothermal system itself (i.e., source of H_2S, energy, sulfide surfaces, etc.). Entropy production itself cannot be sufficient for self-organization; witness the example of the irreversible conversion of solar radiation to heat at an abiotic surface with an albedo of less than one.

However, it is precisely on planetary surfaces that Morowitz (1989, 1992) predicted "photosynthetically mediated molecular ordering" if not life. This follows from his proof, already mentioned, that systems cycling matter must occur in such nonequilibrium conditions.

Are there some law-governed patterns common to organisms, ecosystems, and the biosphere as a whole, given the state of theory to date? The possible predictive power of the entropy formulation is unclear at this stage; would biospheres in general evolve like the terrestrial biosphere? This is a subject discussed in some detail by Boston and Thompson (1991), although not from a thermodynamic perspective.

As Lesins (1991) pointed out, the total living biomass is far too small to make a significant impact through its irreversible chemical reactions on the total energy dissipated at the surface, by virtue of the radiative fluxes. The potential role of biota in affecting the Earth's albedo was raised by Lovelock both in his original work (Lovelock 1979) and later in the Daisyworld models (Watson and Lovelock 1983). Ulanowicz and Hannon (1987) proposed that living systems generate more entropy than their abiotic environment by altering the albedo (lower at short wavelengths) and emissivity (higher at longer wavelengths). If they are right, then the progressive colonization and increase of biotic productivity on land (McMenamin and McMenamin 1993) should have resulted in a progressive increase in entropic flow from the land surface, from the albedo effect alone. However, there is no clear evidence for a significant biotic albedo effect affecting global temperature over geologic time (the albedos of vegetated land overlap with bare ground and soil; see Sellers 1974). Thus, as we have argued above, the cooling trend of the Earth surface since the origin of life is largely a result of the biota's role in affecting atmospheric composition. Has this pattern of biospheric and climatic evolution been contingent on some early mutation, with dramatically different outcomes (e.g., a strong albedo effect mediated by biota as in Lovelock's Daisyworld models), equally probable if the evolutionary process was "run again"? Or is the pattern a product of quasi-deterministic constraints on evolution, likely to be repeated again and again on Earth-like planets around sunlike stars (e.g., as argued in Schwartzman and Volk 1991b)?

If the biotic enhancement of weathering is now high, and has been rising ever since the origin of life, the self-organization of the biosphere has been geophysiological. Biospheric self-organization has increased with the progressive colonization of the continents and evolutionary developments in the land biota, as a result of surface cooling arising from biotic enhancement of weathering, the equilibrium temperature/pCO_2 of the Urey reaction being the attractor state.

With future solar luminosity increases, and ignoring the possibility of anthropogenic effects, the biospheric capacity for climatic regulation will decrease, leading to the ending of self- organization some 2 billion years from

now. The Earth's surface will then approach chemical equilibrium with respect to the carbonate-silicate cycle. If the self-organization of the Earth's surface system is purely geochemical (inorganic) it will likewise end at the same
time that the surface temperature converges on the equilibrium temperature.

Homeostasis Redux

In a radical return to homeostatic Gaia, Markos (1995) postulates that the
homeostasis of the biosphere (relative constancy of temperature, pH, etc., at
a specified time scale) is maintained by an information acquisition system
interpreting "spatiotemporal cues, based on previous history and experience
of the system." Markos postulated that Gaia "reads" its internal environment
and in turn adjusts it to maintain homeostasis. Information mediators within
Gaia include its ultrastructure, diffusible signals, and gene flux (transfer between organisms mainly via viruses). As such, Gaia is a superorganism.

A similar concept was proposed by Williams (1996), who claimed that the
biogeochemical cycles of elements like nitrogen represent a Gaian emergence of a global molecular machinery. For Williams the sensitivity of a biological process to a given environmental component gives it a potential regulatory role in global metabolism (his Gaia). For example, he postulated that
the molecular mechanisms coded in the genome control the coupled nitro-
gen/carbon cycle. However, he conceded that global homeostasis may not
be a consequence of such "bottom-up" expression of molecular control systems.

In a fascinating discussion, Volk (1998) introduced the concept of a me-
tabolizing Gaia, with its parts consisting of kingdoms, cycles, pools, and so
forth, depending on the perspective of the observer. He suggested that the
fundamental "actors" in this metabolism are biochemical guilds, such as nitrogen fixers, and respirers, which cut across divisions such as kingdoms.
One of the results of the activity of these interacting, complementary guilds
is the extraordinarily high efficiency of organic matter recycling in the ocean.
Apparently, the heterotrophic bacterial guild ferociously attacks organic matter in all its forms, including living organisms (Azam 1998). In Volk's interpretation, Gaia is not a living organism, nor does it or its parts necessarily
remain at homeostasis, but it has a metabolism, a geophysiology.

Smolin (1997) suggested the following mechanism to achieve stability of the biosphere:

> During a very large avalanche of mutations and extinctions, there will be selective advantage for species that can regulate, and hence stabilize the biosphere through the byproducts of its metabolic processes. When such a species arises it can contribute to the stability of the whole system. By doing so it may bring to an end an unstable period, in which the rate of extinctions has been very high; and by doing that it contributes to its own survival (p. 150).

A similar scenario has been suggested for the evolution of early microbial communities facing nutrient shortages (Barlow and Volk 1992b). It has been previously argued here that homeostasis has limits, that homeorrhesis, i.e., a series of steady states, is probably a more accurate description of long-term biospheric behavior. Whether or not Markos and Williams' radical reinterpretation of homeostatic Gaia is verified in the future, this concept is likely to generate fruitful research if taken as seriously as it surely deserves. At the least, the genome of organisms, especially within the higher kingdoms of Plantae, Metazoa, and Fungi, is likely to encode the history of the biogeochemistry of the biosphere since the origin of life, if we were only clever enough to read it more fully. Such major events as the secular decline in temperature and atmospheric pCO_2, and the rise of atmospheric pO_2, should be recorded in the higher eucaryote genome because the biochemistry of these extant organisms is the cumulative heritage of adaptations to these events over geologic time.

As already discussed in chapter 8, the history of endosymbiogenesis appears to be the key to the emergence of eucaryotes, cells with nuclei and aerobic forms with mitochondria, as well as the kingdoms of Eukarya. The origin of cells with mitochondria was plausibly linked to two necessary conditions: the rise of pO_2 in microhabitats and the decrease of ambient temperature below the upper limit for the stability of this cell type. As Dyer and Obar (1994) suggested, the predecessor of the first mitochondria was likely an aerobic bacteria (living model being *Paracoccus*) that had evolved the ability to both detoxify oxygen and utilize it in a new metabolism, namely respiration. They proposed that the internalization of this symbiont into the aerobic eucaryote cell was driven by the utilization of host waste (e.g., acetic acid) by the respirating bacteria.

An alternative scenario for the origin of eucaryotes was proposed by Martin and Muller (1998) (see discussion by Doolittle 1998a). Their hypothesis also invokes an endosymbiogenetic origin from two cells: an autotrophic Archea and a heterotrophic Eubacteria. They argue that the latter respired hydrogen and carbon dioxide, which was used by the methanogenic archean partner as a source of energy and carbon. As in Margulis' and Dyer and Obar's model, the Eubacteria ultimately ended up as a mitochondria in the eucaryotic cell, but first going through the complementary association described above without free oxygen.

The linkage of the origin of other eucaryotic organelles and metabolism to changes in environmental conditions remains somewhat obscure. For example, a necessary condition for Metazoa may have been the biosynthesis of the structural protein collagen, which requires significant levels of free oxygen (Towe 1970, 1981). As previously argued, such levels were likely present even in Archean microenvironments, such as within stromatolitic mats, so invoking pO_2 as the constraint holding back an earlier emergence of Metazoa is implausible. However, it is likely that higher levels of atmospheric oxygen were indeed necessary for larger Metazoa.

Thus, the historical biogeochemistry of the biosphere should be derivable from three sources: the geochemistry of the sedimentary record, the fossil history of life, and the living genome itself.

The Self-organizing Biosphere: Theses on Biospheric Evolution

For the strict neodarwinian, the biosphere cannot evolve in the sense that the biota evolves. However, it may be fruitful to explore the consequences of considering the analogues of genotype and phenotype in the biosphere itself, indeed its evolution in a different sense from the usual definition of evolution for transbiological systems (e.g., superficial descriptive social and cultural evolution). Let us be clear that use of the terms "genotype" and "phenotype" for the biosphere in the discussion that follows should not be taken in the technical sense the terms are used for organisms. Here are my six theses on biospheric evolution:

1. The biosphere is a complex adaptive system, adapting to changing external abiotic constraints (solar luminosity, volcanic outgassing rate, production rate, mass/area of continental crust) (see remarks on Gaia by Barlow and

Waldrop 1994, p. 217), but also self-adapting (e.g., creating new biogeo-chemical subcycles, new steady states as a consequence of biotic, atmo-spheric, and crustal evolution) and self-selecting (destroying and creating environments and ecosystems, e.g., thermophile and oxygen catastrophes of surface ecosystems, which limited and altered the surface environments of thermophiles and anaerobes, respectively).

2. The biosphere is a self-organizing complex whole. The interpenetra-tion of its parts and its whole includes the nonlinear interaction of the parts, its network of positive and negative feedbacks, the continual reshaping/re-articulation of the parts by the whole, the history of the whole recorded in its parts, transients, and steady-states. The biosphere is a complex totality ("it is a whole whose unity, far from being the expressive or 'spiritual' unity of Leibniz's or Hegel's whole, is constituted by a certain type of *complexity,* the unity of a *structured whole* containing what can be called levels or instances which are distinct and 'relatively autonomous', and co-exist within this com-plex structural unity, articulated with one another according to specific de-terminations . . ." (Althusser 1970; see Schwartzman 1975). As Levins and Lewontin (1985) put it:

> The interpenetration of parts and wholes is a consequence of the inter-changeability of subject and object, of cause and effect . . . .. Because elements recreate each other by interacting and are recreated by the wholes of which they are parts, change is a characteristic of all systems and all aspects of all systems.

The whole here is the biosphere; the parts include ocean, upper crust, lower atmosphere, surface ecosystems, and fresh water. The failure to recognize the dialectical interaction of the whole and parts leads to the errors of "holism" and reductionism. An example of relevant holism is the assertion that the biosphere itself is a living organism. Examples of reductionism include claims that there are no emergent properties of the biosphere, that an understanding of either molecular metabolism or physics and chemistry is sufficient to pre-dict global biogeochemistry from first principles.

It is one thing to assert that the biosphere is a self-organizing complex whole; it is another to ascend from the abstract to the concrete, i.e., to work out on the basis of empirical evidence the complex dynamics of such a whole and the interaction of its parts. This is a research program that is likely to occupy several more generations of scientists.

3. The genotype of the biosphere is its material inheritance, the sum total of all its parts, embodying its history (genetically coded or preserved),
boundary conditions, and structure. Let us assume that the biota evolves in
the commonly accepted mechanisms of the neodarwinian paradigm, plus endosymbiosis and possibly other mechanisms. In contrast, the biosphere evolves
in a Lamarckian mode (acquired characteristics are inherited, constituting
its genome). The phenotype of the biosphere is at any time its activity, its
metabolism, the character of its biogeochemical cycles (fluxes, steady states,
temperatures, partial pressures, etc.). The genotype of the biosphere is the
cumulative product of its phenotypes since its birth (the origin of life).

4. The history of the phenotypes of the biosphere is recorded in its parts,
that is, fossils, geochemistry of sedimentary and metasedimentary rocks (and
possibly even by igneous rocks in so far their chemistry and isotopic composition have been affected by subduction of biogenic sediments), as well as in
the genome of the biota [i.e., its enzymatic systems for biogeochemical cycling, the historical geochemistry recorded in its phylogeny, its history of
endosymbiosis, e.g., hypersea (McMenamin and McMenamin 1994)]. This
is an example of the history of the whole reflected back into and recorded in
its parts. It also creates the potentiality of rapid reemergence and speciation
following catastrophe (e.g., big impacts). Of course, the phenotypic activity
of the biota is a component of the biosphere's phenotype (Vernadsky's "life
as a geological force").

5. The Gaian character of biospheric evolution includes the tight coupling
of the abiotic and biotic components, its self-regulation. Its Vernadskian
character consists of the progressive changes in biospheric history (cooling,
increase in productivity of biota, invasion of the crust, hypersea, the cumulative increase in diversity of ecological niches over time, etc.).

6. The evolution of the biosphere self-selects a pattern of biotic evolution
that is quasi-deterministic in its main contours given similar initial conditions
and external perturbations (e.g., impact history). Hence, in the future, with
the exploration of space, the theory of comparative biospheres. This view is
very Vernadskian: "The non-accidental character of structure and function
of the biosphere is a constant theme in Vernadsky's work. . . . The ideas of
stochastic variation, undirectedness, and unpredictability were alien views"
(Ghilarov 1995).

A heuristic approach to developing a theory of the biosphere was described by Yates (1987b):

At its largest scale, that of the terrestrial biosphere, biological processes will yield to statistical thermodynamic modeling of the overall biosphere-lithosphere-hydrosphere-atmosphere ecological system in which each component shapes the others and is in turn shaped by them [geophysiology]. The global, structural stability of the whole system will again be seen to express six main processes: activation or inhibition, cooperation or competition, chemical complementarities and broken symmetries—all features that can be broadly formulated mathematically. The specifics are, as always, the hard part and progress will be, for a long time to come, case-by-case (p. 639).

These concepts of stability are taken from physics and chemistry (e.g., enzymatic inhibition, the broken symmetry of phase transition, self-organization of far from equilibrium systems).

Kauffman (1995) suggested that, as characteristics of their self-organization, local ecosystems evolve to the subcritical-supracritical boundary, whereas the biosphere as a whole is supracritical, evolving inexorably to greater molecular diversity and complexity. In general, supracriticality refers to the state of everexpanding complexity, the self-organization of new ordered systems, whereas subcriticality refers to a state of relatively stable subsystem diversity (e.g., molecular and genomic). However, on a geologic time scale the biosphere, ignoring the possibility of global anthropogenic engineering that may postpone its demise, has a finite lifetime constrained by the sun's inexorable increase of luminosity (see earlier discussion). Thus, future evolution will return the biosphere to the subcritical regime, the "garden of thermophiles," before its extinction, assuming that Kauffman's description is correct.

Appendix: Calculation of Effective, Abiotic, and Equilibrium Temperatures

The effective black body radiation temperature, T_e, of the Earth was assumed to vary with age (t in Ga) as follows (see Kasting and Grinspoon 1990):

$$T_e = 255/(1 + 0.087t)^{0.25}$$

This expression assumes a constant planetary albedo.

Abiotic temperatures were computed from an assumed $B = 100$ (see

Schwartzman and Volk 1991a); the biotic enhancement of weathering on the present Earth is 100.

Equilibrium temperatures were computed from thermodynamic data for the Urey reaction:

$$CaCO_3 + SiO_2 = CaSiO_3 + CO_2; \Delta G_o = 9984 \text{ cal/mole}$$

$K = e^{-(4992/T)} = pCO_2$, the partial pressure of carbon dioxide (in atm) in the atmosphere/ocean system.

With a greenhouse function that takes into account the variation in solar luminosity, both the temperature and carbon dioxide level are soluble at any time. The equilibrium temperatures and corresponding atmospheric carbon dioxide levels represent hypothetical states at low temperatures because of the very slow kinetics of the abiotic solid/gas reaction at these conditions. The computed equilibrium temperatures are a first approximation only of what on a real Earth would be more complex equilibria involving CaMgFe silicates and carbonates.

We used the greenhouse function provided by Caldeira and Kasting (1992b) because it gives more plausible temperatures for the low pCO_2 range than the Walker et al. (1981) function (Kasting, personal communication):

$T = T_e + \Delta T$, where T is the mean global surface temperature (°K).

$\Delta T = 815.17 + (4.895 \times 10^7)T^{-2} - (3.9787 \times 10^5)T^{-1}$
$\qquad - 6.7084y^{-2} + 73.221y^{-1} - 30{,}882\,T^{-1}y^{-1}$,

where $y = \log pCO_2$ (in bars).

We assumed a constant planetary albedo, using the above expression for the variation of T_e (Caldeira and Kasting assumed a dependence of albedo on temperature). Our computed temperatures are only roughly comparable with those of Caldeira and Kasting owing to model assumptions; for example, Caldeira and Kasting's biotic temperatures were deliberately maximized with atmospheric pCO_2 held at 10^{-6} bars at times greater than 1 billion years in the future.

Implications to bioastronomy (SETI): habitability of terrestrial planets. A theory of cosmic biospheres? The timing of emergence of complex life and intelligence. Future detection of alien biospheres (e.g., spectral identification of ozone indicating the presence of oxygen, and hence photosynthesis on extrasolar planet?).

Implications to Bioastronomy

Prominent evolutionary biologists have for the most part been hostile to the Search for Extraterrestrial Intelligence (SETI). Their arguments center on the purported "incredible improbability of genuine intelligence emerging" (Mayr 1985; see Simpson's now classic statement, 1964). Tipler (1980), the great opponent of SETI, puts it as follows:

> The present-day theory of evolution, the Modern Synthesis, stresses the great contingency of all branches of the evolutionary tree: those possible branches which could terminate in intelligent life are extremely small in number when compared with the total number of branches.

Some evolutionary biologists are not so sure. Gould (1985) says:

> But does intelligence lie within the class of phenomena too complex and historically conditioned for repetition? I do not think its uniqueness on earth specifies such a conclusion. Perhaps, in another form on another world, intelligence would be as easy to evolve as flight on ours.

Bieri (1964), in a rejoinder to Simpson, went even further in contending that humanoids would be a likely outcome of parallel evolutionary convergence on other presumably terrestrial planets. Bieri cited the examples of convergence well known in evolution such as between dolphins and sharks, as well as the advantageous arrangements found in hominoids (mouth at one end, anus at the other).

A deep question raised by this controversy is the repeatability of patterns of evolution. If evolution of life on Earth was "run again," how similar would the history of life be to the one we know, assuming the same initial conditions and historical abiotic constraints? We can all agree on the great improbability of the exact repetition of terrestrial history on other Earth-like planets in the galaxy. Yet, to what extent is the general pattern of terrestrial evolution (e.g., procaryotes, eucaryotes, Metazoa, intelligence) "forced" by strong tendencies for life to self-organize in preferred directions? (See discussion by Schwartzman and Rickard 1988.) For example, Russell (1981) has argued that progressive encephalization occurred in a fairly regular pattern over the past 600 million years. This pattern may simply be a result of random evolution from a small brain relative to body weight, to a large brain, given the increasing number of extreme values likely with time as the branches of evolution diverge (see Gould 1996, on Cope's Rule, regarding an analogous trend of larger animal size with evolution).

However, the emergence of oxygenic photosynthesis by cyanobacteria, the endosymbiogenesis of bacteria to form the eucaryotic cell, and the critical inventions necessary for the emergence of Metazoa (e.g., biochemistry of embryonic development, structural proteins) may be another case altogether, despite the canonical views of such evolutionary biologists as Mayr, as stated in 1961—"probably nothing in biology is less predictable than the future course of evolution"—and 1985—the emergence of eucaryotes by symbiosis was a "most improbable event." If future research confirms that these inventions occurred soon after, on a geological time scale, all the necessary environmental conditions were set, then a certain inevitability is supported. Temperature has been identified here as the last limiting factor holding up the emergence of major organismal groups. This would explain the puzzling lag times in evolution, evidence Gould (1989) has used to support the canonical view of evolutionary biology that historical contingency is central. In the case of oxygenic photosynthesis, is it not plausible that the invention of this metabolism is forced by the very presence of liquid water, carbon dioxide, and the solar flux? Similarly, the emergence of eucaryotes, as previously argued, may have been the highly likely consequence of the coupling of the complementary metabolisms of two or more procaryotes.

Recently, it has been suggested that spontaneous self-organization has played a critical role in evolutionary change (Waldrop 1990; Kauffman 1993, 1995). Thus, a theory of evolving self-organizing systems may yet emerge that predicts finite patterns of biospheric and biotic evolution on terrestrial planets within the habitable zone (HZ). Vernadsky (1945) put it as follows: "the biosphere . . . is being revealed as a planetary phenomenon of cosmic character." (He went on to say that Venus and Mars had life "beyond doubt.")

Habitability of the Earth

Habitability for life is defined by the constraints of temperature, pH, and other physical and chemical parameters. The presently estimated upper temperature limit for life (for viable growth) is about 125°C, the limit for presently known hyperthermophiles. Life at this temperature is only possible at pressures sufficient to keep water liquid. Some believe hyperthermophiles may be discovered that are viable at a somewhat higher temperature (150°C), the apparent upper limit for protein stability sufficient to make metabolism possible (Brock et al. 1994). The lower temperature limit for growth is a few degrees below 0°C (saline water is still liquid). The origin of life and its persistence as thermophiles below ice cover in hydrothermal vents deeper in the crust might occur on an ice-covered Earth; one could also imagine the emergence of lower temperature forms, at least microbes.

These limits define the upper and lower temperature boundaries for "life as we know it" (one could speculate about exotic biochemistries in alien biospheres; see Feinberg and Shapiro 1980). The upper habitability limit for life was plausibly established between the accretion of the Earth some 4.5 billion years ago and the end of the late bombardment of Earth by leftover debris from planetary accretion some 3.7 to 3.8 billion years ago.

Recently, several molecular biologists suggested that the origin of life took place around 4.2 Ga, based on the considerable complexity evident in the earliest fossil record (i.e., probable cyanobacteria by 3.5 Ga), as well as detailed arguments related to molecular phylogeny (see chapter 8). Large impacts may have even sterilized the surface (Oberbeck and Mancinelli 1994), driving the first life back into hydrothermal vents in the crust, the birthplace of hyperthermophiles (see chapter 8). In any case, the abiotic history of the solar system created the first window of opportunity for life to emerge. Most origin of life researchers think this happened very fast once survival of the earliest life form was possible. Because the luminosity of the sun steadily increased according to the standard theory (see chapter 3), if the Earth at 3.8 Ga contained a pressure-cooker atmosphere, with a surface temperature around 85°C, then were it not for the removal of atmospheric carbon dioxide by the carbonate-silicate cycle the temperature might have exceeded the habitability limit early in Earth history. Conversely, as Caldeira and Kasting (1992a) have shown, were it not for very warm conditions early on, the Earth might have plunged into an irreversible global ice age as a result of carbon dioxide cloud condensation raising the planetary albedo (although this mechanism has been recently questioned; Forget and Pierrehumbert 1997).

To jump some 6 billion years forward (1–2 billion years from now), the upper habitability limit will again be reached as atmospheric carbon dioxide level drops to zero, with a water greenhouse inexorably raising surface temperature (Caldeira and Kasting 1992b). Thus, 6 billion years of habitability with respect to surface temperature is:

1. A lucky accident [invoking the anthropic principle here, as Doolittle (1981) did, we are here to recognize this history, because "only a world which behaved as *if* Gaia did exist is observable because only such a world can produce observers") or
2. A product of deterministic self-regulation of the Earth's surface system.

This second alternative might claim that the silicate-carbonate biogeochemical cycle, the fundamental biospheric self-regulator of temperature, made possible continuous habitability from 4.2 Ga to the future end of the biosphere (a caveat is in order here: some researchers have suggested that at least surface life was reinvented several times in the early Precambrian). However, it is important to distinguish here the habitability for life, and habitability for low temperature life, or at least for complex life. The abiotic boundary conditions may well have guaranteed continuous habitability for thermophiles for 6 billion years, but complex life may have required biotically mediated cooling via the geophysiological climatic stabilizer.

A lucky accident explanation is ultimately testable by the discovery of alien biospheres and their survival probabilities and dynamics. But an under-standing of the latter for our own biosphere could conceivably rule out the lucky accident, just as origin of life research now points to its near inevitabil-ity as a phenomenon of self-organization (Kauffman 1993, 1995).

Habitability of Terrestrial Planets

Although Dole (1964) focused on planetary habitability for human life in his pioneering study, most discussions since that time have defined the HZ (CHZ is the continuously habitable zone) by the presence of liquid water alone (e.g., Kasting, Whitmire, and Reynolds 1993). The HZ in a planetary system is thus that space around the star in which liquid water is stable on planetary surfaces. This definition of habitability is conservative because it avoids the issue of habitable interiors of planets and moons, even without atmospheres (Gold 1992) and habitable moons around inhabitable planets (Williams et al. 1997; Chyba 1997).

A liquid water HZ is compatible with terrestrial life forms ranging from hyperthermophilic microbes with upper temperature limits for growth up to about 120°C (water is liquid under pressure) to algae growing in water films near 0°C in snow fields. Eucaryotes with mitochondria—henceforth to be called complex life, with apologies to Lynn Margulis who has empha-sized the complexity of procaryotes—have an upper temperature limit for viable growth of about 60°C, a limit apparently determined by the thermola-bility of the mitochondrial membrane (Brock and Madigan 1991). Hence, the width of the HZ for complex life on Earth-like planets could be signifi-cantly smaller than that for procaryotes (because thermophilic procaryotes do well above that temperature) assuming that ambient temperatures remain above 60°C for the lifetime of the biosphere. The relevance of such a conjec-ture is of course dependent on the cosmic replication of the general pattern of evolution of life found here on Earth. The consequence of this largely ignored fact will be discussed here in the context of the evolution of the Earth's biosphere, the integrated system composed of the biota and the space/time it inhabits and influences, namely the atmosphere, oceans, and solid crust down several kilometers, as Gold (1992) has described. We have argued that biotically mediated cooling of the surface of Earth-like plan-ets is necessary for the origin and continued evolution of complex life

(Schwartzman and Volk 1991b; Schwartzman and Shore 1996). Temperature on an abiotic planet would remain above the 60°C limit for complex life.

Here it is argued that the statement "no life, no complex life" has a non-tautological interpretation. It is assumed that the emergence of relatively simple life forms is almost inevitable on Earth-like planets. Such organisms are remarkably robust. Complex life, however, requires a more restrictive set of physical conditions, specifically, lower temperatures. Although any non-panspermic (i.e., not "seeded" from extraterrestrial sources) origin of complex life may require the prior existence of primitive life, the former may additionally require the latter in a biospheric sense, that is, via biotically mediated cooling along the lines we are proposing. Biotically mediated cooling may substantially alter the size of the HZ around stars for the emergence of complex life.

Further Implications to the CHZ for Complex Life

The surface temperature of a planetary surface is a function of the recepted solar flux, planetary albedo, and levels of greenhouse gas. The inner boundary of the CHZ for complex life is determined by a solar flux relative to the present, S_{eff}, equal to 1.13, corresponding to a surface temperature (T_s) of 60°C at $pCO_2 = 0$ (Caldeira and Kasting 1992b), because this is the limit to biotically mediated cooling (except perhaps cooling arising from a Love-lockian biotic effect on the planetary albedo). Thus, the inner boundary of the CHZ is close to that for water loss in a few hundred million years ($S_{eff} = 1.10$; Kasting et al. 1993).

The results of an abiotic Earth/sun system is shown in figure 10-1, computed assuming our preferred model b for volcanic outgassing and land area as a function of time (see chapter 8 Appendix). This CHZ is limited by the boundaries for water loss (inner, $D = 0.95$ astronomical units [AU]) and first CO_2 condensation (outer, $D = 1.15$) (Kasting, Whitmire, and Reynolds 1993). The scenario of cooling from dry ice clouds in CO_2-rich atmosphere has been challenged (Forget and Pierrehumbert 1997). Note that t_c, the time (in billions of years) since the emergence of complex life ($T_s \le 60°C$), varies inversely with the assumed B, the present biotic enhancement of weathering factor, for a given D, the distance of the planet from the sun (AU) (figure 10-2); i.e., the higher the assumed value of B on a biotic Earth, the higher

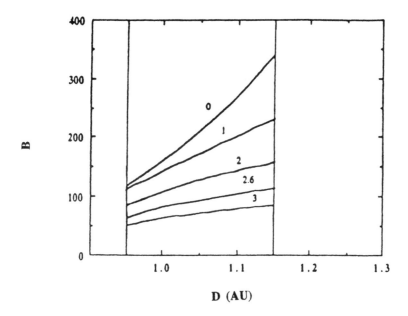

FIGURE IO-I.

Biotic enhancement of weathering factor B versus D, for an Earth-sun model, with the distance of the Earth from the sun, D, in AU (preferred model b results). Numbers on curves are t_c, the time (in billions of years) since the emergence of complex life ($T_s \leq 60°C$). The inner ($D = 0.95$) and conservative outer ($D = 1.15$) boundaries of the liquid water CHZ are shown.

the surface temperature, and the longer it would persist above 60°C, the upper temperature limit of complex life, on an abiotic Earth surface (t_c becomes smaller), without the benefit of biotic cooling. As D increases, it becomes easier to cool the planet (t_c increases) for a given B.

Figure 10-3 illustrates the relationship between D and t_c for given assumed B values. For $t_c \geq 2$ billion years (i.e., eucaryotes emerged at least 2 billion years ago) and B is assumed greater than 160, $D > 1.15$, the conservative outer limit of the CHZ; that is, on an abiotic Earth-like planet ($D = 1$) around a sunlike star, the surface temperature remains above 60°C over the whole CHZ. In conclusion, the biotically mediated cooling arising from biotically enhanced chemical weathering is postulated to have been crucial to the emergence of complex life on Earth.

This model of biotically mediated cooling is now applied to other stars. We have suggested that biotically mediated surface cooling may have signifi-

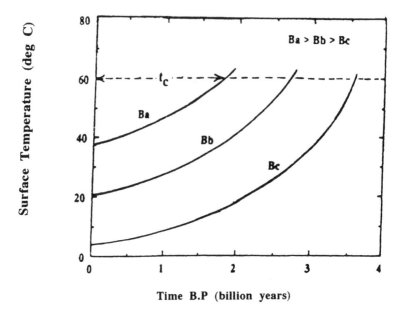

FIGURE 10-2.

Surface temperature versus time for a given D; The B's are biotic enhancement factors; t_c is indicated.

cantly enhanced the possibility for the evolution of complex life and intelligence on Earth-like planets around sunlike stars (Schwartzman and Volk 1991b; Schwartzman 1994b). If surface temperature is indeed a critical constraint on the emergence of complex life, then the time needed for its appearance dated from the origin of life (Δt_c) may be shorter on biospheres of terrestrial planets within the HZ, which cool more quickly than the Earth (conversely for slower cooling). Shklovskii and Sagan (1966) proposed the assumption of mediocrity, that is, the time it has taken to evolve life and intelligence on Earth is average for habitable planets in the galaxy. However, this postulate may be misleading as a guide to the evolution of alien biospheres. Complex life and intelligence may be more or less common in the galaxy than deduced by simply extrapolating the time frame of evolutionary change on the Earth. A better understanding of the evolution of our own biosphere could inform a predictive theory of alien biospheres. In particular, for terrestrial planets, Δt_c and Δt_i (time needed for intelligence to emerge) may prove to be an explicit function of planetary mass, distance from its star, and stellar mass.

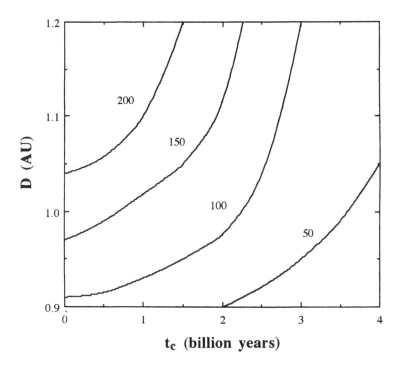

FIGURE IO-3.

D versus t_c for an Earth-sun model; numbers on curves are B values (preferred model b results). The inner ($D = 0.95$) and conservative outer ($D = 1.15$) boundaries of the liquid water CHZ are shown.

For an Earth-like planet within a liquid water CHZ scaled to our sun (Kasting et al. 1993), the following relationships are predicted, assuming all other variables are held constant:

For $M_{star} > M_{sun}$; then $\Delta t_{c,i\ star} > \Delta t_{c,i\ sun}$ and conversely, for $M_{star} < M_{sun}$; then $\Delta t_{c,i\ star} < \Delta t_{c,i\ sun}$ (M is the mass of the star)

These results follow from the rate of luminosity increase with time, dL/dt, relative to that of the sun; biotically mediated cooling must fight dL/dt. For an Earth-like planet in the CHZ around an F0 star[*], there is "double

---

[*]Stars on the main sequence are classified by their luminosity (and mass) in decreasing order: OBAFGKM, with each letter having 10 subclasses (0 to 9). The Sun is a G2 star.

TABLE 10-1.

Some of the Possible Variables Affecting the Time Required for the Emergence of Complex Life Intelligence ($\Delta t_{c,i}$) on Earth-like Planets Around Other Stars[a]

Variable (change from past to present time, relative to Earth-Sun)	$\Delta t_{c,i}$
Luminosity ↑	↑ $M_{star} > M_{Sun}$
Volcanic outgassing rate ↓	↓ No plate tectonics, lower K, U
Continental land area growth rate ↑	↓ Smaller initial oceans
Biotic enhancement of weathering ↓	↑ Different evolutionary pattern
Impact size, frequency ↓	↓ Supply of greenhouse gases drops

[a]Scaled to the liquid water CHZ of the Earth-Sun system.

trouble" because of the star's short lifetime on the main sequence (about 3 billion years). These results suggest that complex life should emerge and evolve faster around stars less massive than and more numerous than sunlike stars in the galaxy. This news will be welcomed by the SETI optimist camp. However, differences in other variables from Earth's unique history may make Earth-like planets around say K or M stars less likely to harbor complex life (table 10-1).

We are likely facing the prospect of comparative cosmogony, with the discovery of hundreds, if not thousands, of extrasolar planetary systems in the coming decades since the discovery of 51 Pegasi, the first apparent extrasolar planet around a solar-type star (see Wetherill 1996; Black 1996; Mayor and Queloz 1995; Mayor et al. 1997; Butler and Marcy 1997). What is not generally appreciated is that this research program may include the discovery of alien biospheres in the first decades of the next century, by the spectral identification of ozone indicating the presence of oxygen, and photosynthesis on extrasolar planets (Mariotti et al. 1997). However, the detection of ozone alone may not be conclusive evidence of a biosphere because traces of oxygen can be generated by abiotic photodissociation of water (see discussion of detection of ozone on two of Saturn's satellites by Noll et al. 1997).

Implications to Evolution on Mars

In a final note of wild speculation, we predicted that complex life emerged earlier on Mars than on Earth because the Martian surface cooled faster than that of Earth (Schwartzman and Shore 1996). This prediction does not require that biotically mediated cooling did occur on Mars, or even that Mars was in the HZ defined by persistent liquid water at its surface with temperatures necessarily above its freezing point (Kasting et al. 1993) in its early history, an issue still under debate. We only assume that the origin of life did occur on Mars, at the same time as on Earth, that habitable environments did exist [e.g., in ice-covered lakes (McKay and Davis 1991) or in hot springs (Walter and Des Marais 1993)], and that life was not aborted in say the first billion years (e.g., by impacts). An extant biostratigraphic record in the Martian crust will test this admittedly outrageous hypothesis. We may find out sooner than we think. In the interest of brevity, we will skip a discussion of the possible fossils in the Martian meteorite.

Could life have persisted on Mars to the present, assuming it was once there? If the origin of life occurred in a hydrothermal environment, as it appears was the case on Earth, it is plausible that a presently living Martian microbial biota is derived from an originally thermophilic and hyperthermophilic population. An analogue may be found in terrestrial thermophiles living in arctic, subarctic, and alpine habitats (Boyd et al. 1990). As the Martian surface cooled, thermophiles adapted to soil and endolithic environments, with a population still existing in hot spring–like environments associated with the tail end of Martian volcanism. A presently existing microbial biota has been postulated by Digregorio (1997) based on the admittedly controversial interpretations of the labeled release Viking experiments and other observations by the principal investigators Gilbert Levin and Patricia Ann Straat.

Conclusion

We have suggested that biotically mediated surface cooling may have significantly enhanced the possibility for the evolution of complex life and intelligence on Earth-like planets around sunlike stars (Schwartzman and Volk

1991b). If surface temperature is indeed a critical constraint on the emergence of complex life, then the time needed for its appearance dated from the origin of life (Δt_c) may be shorter on biospheres of terrestrial planets within the HZ, which cool more quickly than the Earth (conversely for slower cooling). In fact, the assumption of mediocrity (Shklovskii and Sagan 1966) may be misleading as a guide to the evolution of alien biospheres. Complex life and intelligence may be more or less common in the galaxy than deduced by simply extrapolating the time frame of evolutionary change on the Earth. However, a better understanding of the evolution of our own biosphere could inform a predictive theory of alien biospheres. For terrestrial planets, Δt_c and Δt_i (time needed for intelligence to emerge) may prove to be an explicit function of planetary mass, distance from its star, and stellar mass.

The width of the HZ for complex life may be substantially smaller than that for the appearance of biota, constrained by the presence of liquid water. The consequences of this largely ignored fact have important implications to the SETI research program. A geophysiological climatic mechanism has been outlined in this book. Perhaps alien biospheres have analogous geophysiologies. Surface temperature history on terrestrial planets may be critical to the time needed to evolve complex life and intelligence.

Biotic colonization of continents and resultant enhancement of weathering, first by thermophilic bacteria, led to lower steady-state atmospheric carbon dioxide levels and surface cooling on Earth. Evolution of procaryotes and complex life on terrestrial planets are expected to have similar geochemical and climatic consequences. Biotically mediated cooling increases the width of the HZ for the possible occurrence and evolutionary time frame of complex life. For Earth-like planets within the HZ of stars less massive than the sun, the earlier emergence of complex life is expected, all other factors being the same.

The reader is referred to Boston and Thompson (1991) for an interesting discussion of possible scenarios for biospheric evolution. Other astronomical issues affecting habitability include obliquity (tilt of rotation axis) stability of a terrestrial planet, which is likely related to the presence of a moon (Kasting 1996).

:: 11 CONCLUSIONS

Summary of main conclusions. Suggested future directions for research (climatology, geochemistry/geology, geomorphology, field studies, paleontology, biochemis-
try/biophysics/biology).

Summary of Main Conclusions

The salient theses of this book can be summed up as follows:

1. The net effect of land biota is to significantly intensify chemical weathering rates of minerals that lead to the removal of carbon dioxide from the
atmosphere/ocean system, thus leading to a lower atmospheric pCO_2 and
surface temperature of the Earth than would be the case for an abiotic state.
The progressive increase in the biotic enhancement of weathering over geologic time has cooled the surface, given the abiotic constraints of the carbonate-silicate geochemical cycle.

2. The cooling of the Earth's surface, particularly in the first two thirds
of its history, has permitted the emergence of major organismal groups (cyanobacteria, eucaryotes, Metazoa) at such times when the surface temperature corresponded to their biochemically/biophysically determined upper
temperature limit for growth.

3. Thus, the biotic and abiotic components of the biosphere are linked
in a geophysiological feedback loop, which itself changes its characteristics
(evolves) over geologic time.

Are these theses at worst fruitful errors? Only time will tell. Our earlier work has already played a role in stimulating a research program that should provide critical empirical and theoretical tests. To this end, I propose several research directions to extend this now embryonic research program.

Suggested Future Directions for Research

CLIMATOLOGY

General climate models (GCMs), produced from high-speed computer simulations, of a very warm Archean/early Proterozoic climate should be explored for insight into issues of greenhouse responses, latitudinal temperature and precipitation differences, and lapse rates. There have been previous GCMs of the Archean climate, but they have assumed conventionally accepted low global mean temperatures (Henderson-Sellers et al. 1991; Jenkins et al. 1993). Modeling studies are merited of possible excursions from ambient temperatures of a very warm Archean/early Proterozoic as a result of impact, cosmic dust clouds, volcanic pulses, and biotic evolution.

GEOCHEMISTRY/GEOLOGY

A more sophisticated model of the long-term carbon cycle that extends back into the Archean needs to be developed. One major challenge is to better reconstruct the Precambrian surface temperatures and atmospheric compositions (especially pCO_2 and pO_2), as well as the pH and chemistry of the ocean. The oxygen isotopic paleothermometer needs to be improved, with results better understood, perhaps by determinations on smaller samples correlated with other isotopic systems (such as silicon and iron; Beard and Johnson 1999), microtexture, and organic geochemistry of ancient sediments. Laser ablation multiple collector inductively coupled mass spectrometry (Halliday et al. 1997) has shown great promise in the analysis of rocks with diagenetic overprints and microscopic domains of primary climatic records.

New paleothermometers and paleoatmospheric indicators are needed. Rye et al.'s (1995) innovative study on Archean and early Proterozoic paleosols (while I question its conclusions) needs to be followed up. Other aspects of the sedimentary geochemical record, such as phase relations of authigenic minerals and trace element partitioning, probably have promise if only we were clever enough to interpret them. Sedimentary structures such as ripple marks may have potential as indicators of conditions such as salinity and water temperature (Bouguchwal and Southard 1990; Southard and Bouguchwal 1990a, 1990b; Knauth and Grotzinger, personal communication), with the advantage of penetrating the diagenetic/metamorphic veil that has reset other primary systems in the rock.

GEOMORPHOLOGY

The proposal by Molnar and England (1990) that links climate change, weathering, erosion, and isostatic rebound in a feedback system is in itself the beginning of a fruitful research program that promises to enlarge our understanding of both contemporary and ancient climate. The research of Stallard, Dietrich, and others on the interaction of geomorphology and physical/chemical weathering is sure to be extended in several promising directions that are relevant to the long-term carbon cycle. The role of frost wedging as an agent of physical weathering needs to be better understood as a function of global mean surface temperature and topography. The extent of regolith generated on an abiotic Earth and its concomitant increase of potentially reactive mineral surface area needs a closer look by geomorphologists.

FIELD STUDIES OF WEATHERING

The field studies needed to better understand the role of biology in physical and chemical weathering are just beginning. More attention is needed on bare rock, lichen, and bryophyte weathering, both in the field using mini-watersheds (Schwartzman et al. 1997) and in the laboratory using growth chambers. The possible synergism between frost wedging and microbial colonization should be the basis of several graduate theses.

Studies of lichen weathering intended to shed light on the long-term carbon cycle should arguably have the following characteristics (Cochran and Schwartzman 1995):

1. Mafic to ultramafic rock substrates, which have maximum potential weathering rates and, because of their high calcium and magnesium content, contribute disproportionately to the global silicate weathering carbon sink.

2. Long-lived crustose lichens with known growth rates (such as those used in lichenometry).

3. Rock surfaces with known exposure rates (e.g., lava flows, glaciated bedrock, road cuts).

Under these conditions, weathering rates might be inferred from a comparison of rind thickness, chemistry, and mineralogy from lichen-covered and adjacent bare rock surfaces, particularly from the variation of weathering intensity with lichen age.

In addition, the virtually unstudied endolithic microbial communities, ubiquitous even in temperate climates (Hawksworth, personal communication), should be investigated in view of their potential importance in chemical weathering.

PALEONTOLOGY

The possibility that an actual fossil record of Precambrian land biota exists should be taken more seriously by paleontologists (Horodyski and Knauth's 1994 study of an apparent Proterozoic record stands virtually alone). Terrestrial clastic sediments, along with paleosols, might contain such a record. Is it conceivable that an extant sedimentary and fossil record exists of ancient mountain environments?

BIOCHEMISTRY/BIOPHYSICS/BIOLOGY

A fundamental question is the nature of the apparent biophysical/biochemical constraint on the upper temperature limit for growth of procaryotes and the kingdoms of eucaryotes. Heat shock proteins (HSPs) are likely candidates to investigate because of their multifold roles in cells (molecular chaperones, involvement in mitochondrial membrane synthesis, etc.). Nisbet (1995) suggested that HSPs evolved early as a protection against the very high temperatures around hydrothermal vents. I would add that thermophiles probably also needed protection from local low albedo-induced high temperatures on land. The thermostability of HSPs, along with that of organelles and their membranes (mitochondria, nuclei, etc.) and various enzymatic systems (such as involved in blastula formation and the synthesis of collagen, a key structural protein of Metazoa) should be investigated. Are there any living thermophilic amitochondrial eucaryotes that never had mitochondria in their ancestors, with upper temperature limits above the mitochondrial eucary-otes, that might be models of an Archean presence?

Attempts to synthesize and find in nature living models of possible Pre-cambrian land symbioses such as actinolichens (actinobacteria and algae or cyanobacteria) and primitive lichens (e.g., with chytrid mycobionts) should be pursued. Is the anti-lichen model viable for Precambrian land biota?

Finally, an experimental approach to the stabilization and dissolution of a mineral soil by thermophiles should be pursued, which would be relevant to the scenario of early biotic colonization of the land.

:: REFERENCES

Abbott, D. and M. Lyle. 1984. Age of oceanic plates at subduction and volatile recycling. Geophys. *Res Lett* 11: 951–4.

Adamo, P. and P. Violante. 1991. Weathering of volcanic rocks from Mt. Vesuvius
associated with the lichen *Stereocaulon vesuvianum*. *Pedobiologia* 35: 209–17.

Adams, F. D. 1954. *The Birth and Development of the Geological Sciences*. New York:
Dover.

Ahmadjian, V. 1993. *The Lichen Symbiosis*. New York: Wiley.

Alexandre, A., J-D. Meunier, F. Colin, and J-M. Kond. 1997. Plant impact on the
biogeochemical cycle of silica and related weathering processes. *Geochim Cosmochim Acta* 61: 677–82.

Algeo, T. J., R. A. Berner, J. B. Maynard, and S. E. Scheckler. 1995. Late Devonian
oceanic anoxic events and biotic crises: "rooted" in the evolution of vascular land
plants? *GSA Today* 5: 45, 64–6.

Alibert, C. and M. T. McCulloch. 1990. REE and Nd isotope data in BIF from
Hammersley, Western Australia: Implications for the composition of early Proterozoic seawater. *Geological Society of Australia Abstracts* No. 27, p. 2. Canberra, Australia: ICOG 7.

Allard, P., J. Carbonnelle, D. Dajlevic, J. Le Bronec, P. Morel, M. C, Robe, J. M.
Maurenas, R. Faivre-Pierret, D. Martin, J. C. Sabroux, and P. Zettwoog. 1991.
Eruptive and diffuse emissions of CO_2 from Mount Etna. *Nature* 351: 387–91.

Allegre, C. J., J. Gaillardet, L. Levasseur, L. Meynadier, and P. Louvat. 1997. "The
Evolution of Seawater Strontium Isotopes in the Last Hundred Million Years:
Reinterpretation and Consequences for Erosion and Climate Models." In: *Sev-
enth Annual V. M. Goldschmidt Conference*. LPI Contribution No. 921. Houston,
TX: Lunar and Planetary Institute.

Allegre, C. J. and C. Jaupart. 1985, Continental tectonics and continental kinetics. *Earth Planet Sci Lett* 74: 171–86.

Althusser, L. and E. Balibar. 1970. *Reading Capital.* New York: Pantheon.

Amiotte Suchet, P. and J. L. Probst. 1993. Modelling of atmospheric CO_2 consumption by chemical weathering of rocks: Application to the Garonne, Congo and Amazon basins. *Chem Geol* 107: 205–10.

Anderson, S. P., J. I. Drever, and N. F. Humphrey. 1997. Chemical weathering in glacial environments. *Geology* 25: 399–402.

Aoki, I. 1988. Entropy flows and entropy productions in the Earth's surface and in the Earth's atmosphere. *J Phys Soc Jpn* 57: 3262–9.

Aoki, I. 1992. Entropy physiology of swine—a macroscopic viewpoint. *J Theoret Biol* 157: 363–71.

Armstrong, R. L. 1968. A model for Sr and Pb isotopic evolution in a dynamic earth. *Rev Geophys Space Phys* 6: 175–99.

Armstrong, R. L. 1981. Radiogenic isotopes: The case for crustal recycling on a near-steady-state no-continental-growth earth. *Phil Trans R Soc Lond [A]* 301: 433–72.

Armstrong, R. L. 1991. The persistent myth of crustal growth. *Aust J Earth Sci* 38: 613–30.

Arnold, R. W., I. Szabolcs, and V. O. Targulian, eds. 1990. Global Soil Change, the Report of an IIASA-ISSS-UNEP Task Force on the Role of Soils in Global Change. International Institute of Applied System Analysis, Laxenburg, Austria.

Arthur, M. A. and T. J. Fahey. 1993. Controls on soil solution chemistry in a sub-alpine forest in North-Central Colorado. *Soil Sci Soc Am J* 57: 1122–30.

Ayala, F. J., A. Rzhetsky, and F. J. Ayala. 1998. Origin of the metazoan phyla: Molecular clocks confirm paleontological estimates. *Proc Natl Acad Sci U S A* 95: 606–11.

Azam, F. 1998. Microbial control of oceanic carbon flux: The plot thickens. *Science* 280: 694–6.

Bachmann, E. 1904. Zur Frage des Vorkommens von ölführenden Sphäroidzellen bei Flechen. *Berl Dtsch Bot Ges* 22: 44–6. Die Beziehungen der Kieselfechten zu chrem. *Substrat Berl Dtsch Bot Ges* 22: 101–4.

Badger, M. R. and T. J. Andrews. 1987. Co-evolution of Rubisco and CO_2-concentrating mechanisms. *Prog Photosynth Res* 3: 601–9.

Bahcall, J. N., W. F. Huebner, S. H. Lubow, P. D. Parker, and R. H. Ulrich. 1982. Standard solar models and the uncertainties in predicted capture rates of solar neutrinos. *Rev Mod Phys* 54: 767–99.

Bahcall, J. N. and M. H. Pisonneault. 1992. Stamdard solar models, with and without helium diffusion, and the solar neutrino problem. *Rev Mod Phys* 64: 885–926.

Bailes, K. E. 1990. *Science and Russian Culture in an Age of Revolutions. V. I. Vernadsky and His Scientific School, 1863–1945.* Bloomington: Indiana University Press.

Banfield, J. F. and K. H. Nealson, eds. 1997. *Geomicrobiology: Interactions Between Microbes and Minerals. Reviews in Mineralogy 35.* Washington, DC: Mineralogical Society of America.

Barker, W. W. and J. F. Banfield. 1996. Biologically versus inorganically-mediated weathering reactions: Relationships between minerals and extracellular microbial polymers in lithobiotic communities. *Chem Geol* 132: 55–69.

Barker, W. W., J. W. Thomson, J. F. Banfield, W. W. Bailey, and G. A. Haupt. 1994. Biogeochemical weathering of hornblende syenite [Abstract H51A-5]. *EOS Trans Am Geophys Union* 75: 281.

Barker, W. W., S. A. Welch, and J. F. Banfield. 1997. "Biogeochemical Weathering of Silicate Minerals." In: J. F. Banfield and K. H. Nealson, eds., *Geomicrobiology: Interactions Between Microbes and Minerals. Reviews in Mineralogy,* pp. 391–428. Washington, DC: Mineralogical Society of America.

Barlow, C. and M. M. Waldrop. 1994. Worldview extensions of complexity theory, In: C. Barlow, ed., *Evolution Extended,* pp. 212–218. Cambridge, MA: MIT Press.

Barlow, C., and T. Volk. 1992a. Gaia and evolutionary biology. *BioScience* 42: 686–93.

Barlow, C., and T. Volk. 1992b. Open systems living in a closed biosphere: A new paradox for the Gaia debate. *BioSystems* 23: 371–84.

Barns, S. M. and S. Nierzwicki-Bauer. 1997. "Microbial Diversity in Modern Subsurface, Ocean, Surface environments." In: J. F. Banfield, and K. H. Nealson, eds., *Geomicrobiology: Interactions Between Microbes and Minerals. Reviews in Mineralogy 35,* pp. 35–79. Washington, DC: Mineralogical Society of America.

Barron, E. J., W. W. Hay, and S. Thompson. 1989. The hydrologic cycle: A major variable during Earth history. *Palaeogeogr Palaeoclimatol Palaeoecol* (Global Planetary Change Section) 75: 157–74.

Bates, T. S. and P. K. Quinn. 1997. Dimethylsulfide (DMS) in the equatorial Pacific Ocean (1982 to 1996): Evidence of a climate feedback? *Geophys Res Lett* 24: 861–4.

Beard, B. L. and C. M. Johnson. 1999. High-precision iron isotope measurements of terrestrial and lunar materials. *Geochim Cosmochim Acta* (in press).

Beck, R. A., D. W. Burbank, W. J. Sercombe, T. L. Olson, and A. M. Khan. 1995. Organic carbon exhumation and global warming during the early Himalayan collision. *Geology* 23: 387–90.

Beeunas, M. A. and L. P. Knauth. 1985. Preserved stable isotopic signature of sub-aerial diagenesis in the 1.2-b.y. Mescal Limestone, central Arizona: Implications for the timing and development of a terrestrial plant cover. *Geol Soc Am Bull* 96: 737–45.

Berner, E. K. and R. A. Berner. 1987. *The Global Water Cycle*. Englewood Cliffs, NJ: Prentice-Hall.

Berner, R. A. 1971. *Principles of Chemical Sedimentology*. New York: McGraw-Hill.

Berner, R. A. 1989. Biogeochemical cycles of carbon and sulfur and their effect on atmospheric oxygen over Phanerozoic time. *Palaeogeogr Palaeoclimatol Palaeoecol* (Global Planetary Change Section.), 75: 97–122.

Berner, R. A. 1990a. Atmospheric carbon dioxide levels over Phanerozoic time. *Science* 249: 1382–6.

Berner, R. A. 1990b. Global CO_2 degassing and the carbon cycle: Comment on "Cretaceous ocean crust at DSDP sites 417 and 418: carbon uptake from weathering versus loss by magmatic outgassing." *Geochim Cosmochim Acta* 54: 2889–90.

Berner, R. A. 1991. A model for atmospheric CO_2 over Phanerozoic time. *Am J Sci* 291: 339–76.

Berner, R. A. 1992. Weathering, plants, and the long-term carbon cycle. *Geochim Cosmochim Acta* 56: 3225–32.

Berner, R. A. 1993. Paleozoic atmospheric CO_2: Importance of solar radiation and plant evolution. *Science* 261: 68–70.

Berner, R. A. 1994. GEOCARB II: A revised model of atmospheric CO_2 over Phanerozoic time. *Am J Sci* 294: 56–91.

Berner, R. A. 1995a. "Chemical Weathering and Its Effect on Atmospheric CO_2 and Climate." In: A. F. White and S. L. Brantley, eds., *Chemical Weathering Rates of Silicate Minerals. Reviews in Mineralogy*, Vol. 31, pp. 565–83. Washington, DC: Mineralogical Society of America.

Berner, R. A. 1995b. A. G. Hogbom and the development of the concept of the geochemical carbon cycle. *Am J Sci* 295: 491–5.

Berner, R. A. 1997. The rise of plants and their effect on weathering and atmospheric CO_2. *Science* 276: 544–46.

Berner, R. A. and E. J. Barron. 1984. Comments on the BLAG model: Factors affecting atmospheric CO_2 and temperature over the past 100 million years. *Am J Sci* 284: 1183–92.

Berner, R. A. and E. K. Berner. 1997. "Silicate Weathering and Climate." In: W. F. Ruddiman, ed., *Tectonic Uplift and Climate Change*, pp. 353–65. New York: Plenum.

Berner, R. A. and K. Caldeira. 1997. The need for mass balance and feedback in the geochemical carbon cycle. *Geology* 25: 955–6.

Berner, R. A. and A. C. Lasaga. 1989. Modeling the geochemical carbon cycle. *Sci Am* (March): 74–81.

Berner, R. A., A. C. Lasaga, and R. M. Garrels. 1983. The carbonate-silicate geochemical cycle and its effect on atmospheric carbon dioxide over the past 100 million years. *Am J Sci* 283: 641–83.

Berner, R. A. and K. A. Maasch. 1996. Chemical weathering and controls on atmospheric O_2 and CO_2: Fundamental principles were enunciated by J. J. Ebelmen in 1845. *Geochim Cosmochim Acta* 60: 1633–7.

Berner, R. A. and J-L. Rao. 1997. Alkalinity buildup during silicate weathering under a snow cover. *Aquat Geochem* 2: 301–12.

Bieri, R. 1964. Humanoids on other planets? *Am Scientist* 52: 452–8.

Black, D. 1996. Looking for the twilight zone. *Nature* 381: 474–5.

Blum, A. E., D. M. McKnight, and W. B. Lyons. 1997. "Silicate Weathering Rates Along a Stream Channel Draining into Lake Fryxell, Taylor Valley, Antarctica." LPI Contribution No. 921. *Seventh Annual V.M. Goldschmidt Conference,* pp. 29–30. Houston, TX: Lunar and Planetary Institute.

Blum, A. E. and L. L. Stillings. 1995. "Feldspar Dissolution Rates." In: A. F. White and S. L. Brantley, eds., *Chemical Weathering Rates of Silicate Minerals. Reviews in Mineralogy,* Vol. 31, pp. 291–351. Washington, DC: Mineralogical Society of America.

Blum, J. D., C. A. Gazis, A. D. Jacobson, and C. P. Chamberlain. 1998. Carbonate versus silicate weathering in the Raikhot watershed with the High Himalayan Crystalline Series. *Geology* 26: 411–4.

Bormann, B. T., D. Wang, F. H. Bormann, G. Benoit, R. April, and M. C. Snyder. 1998. Rapid, plant-induced weathering in an aggrading experimental ecosystem. *Biogeochemistry* 43: 129–55.

Bormann, F. H., W. B. Bowden, R. S. Pierce, S. P. Hamburg, G. K. Voigt, R. C. Ingersoll, and G. E. Likens. 1987. "The Hubbard Brook Sandbox Experiment." In: R. Jordan, M. E. Gilpin, and J. D. Aber, eds., *Restoration Ecology,* pp. 251–56. New York: Cambridge University Press.

Boston, P. J. and S. L. Thompson. 1991. "Theoretical Microbial and Vegetation Control of Planetary Environments." In: S. H. Schneider and P. J. Boston, eds., *Scientists on Gaia,* pp. 99–117. Cambridge, MA: MIT Press.

Bottomley, D. J., J. Veizer, H. Nielsen, and J. Moczydlowska. 1992. Isotopic composition of disseminated sulfur in Precambrian sedimentary rocks. *Geochim Cosmochim Acta* 56: 3311–22.

Bouguchwal, L. A. and J. B. Southard. 1990. Bed configurations in steady unidirectional water flows. Part 1. Scale model study using fine sands. *J Sed Petrol* 60: 649–57.

Bouthier de la Tour, C., C. Portemer, M. Nadal, K. O. Stetter, P. Forterre, and M. Duguet. 1990. Reverse gyrase, a hallmark of the hyperthermophilic Archaebacteria. *J Bacteriol* 172: 6803–8.

Bowring, S. A. and T. Housh. 1995. The Earth's early evolution. *Science* 269: 1535–40.

Bowring, S. A., I. S. Williams, and W. Compston. 1989. 3.96 Ga gneisses from the Slave province, Northwest Territories, Canada. *Geology* 17: 971–5.

Boyd, W. L., D. R, Onn, and J. W. Boyd. 1990. Therrmophilic bacteria among arctic, subarctic, and alpine habitats. *Arctic Alpine Res* 22: 401–11.

Brady, P. V. 1991. The effect of silicate weathering on global temperature and atmospheric CO_2. *J Geophys Res* 96(B11): 18,101–18,106.

Brady, P. V. 1997. Coupled effects of runoff, temperature, and organic activity on basalt weathering. Geological Society of America Abstracts with Programs 29, No.7: A-86.

Brady, P. V. and S. A. Carroll. 1994. Direct effects of CO_2 and temperature on silicate weathering: Possible implications for climate control. *Geochim Cosmochim Acta* 58: 1853–6.

Brady, P. V. and S. R. Gislason. 1997. Seafloor weathering controls on atmospheric CO_2 and global climate. *Geochim Cosmochim Acta* 61: 965–73.

Brantley, S. L. and Y. Chen. 1995. "Chemical Weathering Rates of Pyroxenes and Amphiboles." In: A. F. White and S. L. Brantley, eds., *Chemical Weathering Rates of Silicate Minerals. Reviews in Mineralogy,* Vol. 31, pp. 119–72. Washington, DC: Mineralogical Society of America.

Brantley, S. L. and K. W. Koepenick. 1995, Measured carbon dioxide emissions from Oldoinyo Lengai and the skewed distribution of passive volcanic fluxes. *Geology* 23: 933–6.

Brantley, S. L. and L. Stillings. 1996. Feldspar dissolution at 25°C and low pH. *Am J Sci* 296: 101–27.

Breyer, J. A., A. B. Busbey, R. E. Hanson, and E. C. Roy III. 1995. Possible new evidence for the origin of metazoans prior to 1 Ga: Sediment-filled tubes from the Mesoproterozoic Allamoore Formation, Trans-Pecos, Texas. *Geology* 23: 269–72.

Bricmont, J. 1996. "Science of Chaos or Chaos in Science?" In: P. R. Gross, N. Levitt, and M. W. Lewis, eds., *The Flight from Science and Reason,* pp. 131–75. New York: New York Academy of Sciences.

Brimhall, G. H., O. A. Chadwick, C. J. Lewis, W. Compston, I. S. Williams, K. J. Danti, W. E. Dietrich, M. E. Power, D. Hendricks, and J. Bratt. 1991. Deformational mass transport and invasive process in soil evolution. *Science* 255: 695–702.

Brock, T. D. 1978. *Thermophilic Microorganisms and Life at High Temperature*. New York: Springer-Verlag.

Brock, T. D. 1986. "Introduction: An Overview of the Thermophiles." In: T. D. Brock, ed., *Thermophiles—General, Molecular and Applied Microbiology*, pp. 1–16. New York: Wiley & Sons.

Brock, T. D. and M. T. Madigan. 1991. *Biology of Microorganisms*, 6th ed. Englewood Cliffs, NJ: Prentice-Hall.

Brock, T. D., M. T. Madigan, J. M. Martinko, and J. Parker. 1994. *Biology of Microorganisms*, 7th ed. Englewood Cliffs, NJ: Prentice-Hall.

Broecker, W. S. and A. Sanyal. 1998. Does atmospheric CO_2 police the rate of chemical weathering? *Global Biogeochem Cycles* 12: 403–8.

Brown, S., L. Margulis, S. Ibarra, and D. Siqueiros. 1985. Desiccation resistance and contamination as mechanisms of gaia. *BioSystems* 17: 337–60.

Brozovic, N., D. W. Burbank, and A. J. Meigs. 1997. Climatic limits on landscape development in northwestern Himalaya. *Science* 276: 571–4.

Burdett, J. W., J. P. Grotzinger, and M. A. Arthur. 1990. Did major changes in the stable-isotope composition of Proterozoic seawater occur? *Geology* 18: 227–30.

Butler, D. R. 1995. *Zoogeomorphology*. New York: Cambridge University Press.

Butler, R. P. and G. W. Marcy. 1997. "The Lick Observatory Planet Search." In: C. B. Cosmovici, S. Bowyer, and D. Werthimer, eds., *Astronomical and Biochemical Origins and the Search for Life in the Universe*, pp. 331–42. Bologna, Italy: Editrice Compositori.

Cai, C., S. Ouyang, Y. Wang, Z. Fang, J. Rong, L. Geng, and X. Li. 1996. An Early Silurian vascular plant. *Nature* 379: 592.

Caldeira, K. 1991. Continental-pelagic carbonate partitioning and the global carbonate-silicate cycle. *Geology* 19: 204–6.

Caldeira, K. 1995. Long-term control of atmospheric carbon dioxide: Low-temperature seafloor alteration or terrestrial silicate-rock weathering? *Am J Sci* 295: 1077–114.

Caldeira, K., M. A. Arthur, R. A. Berner, and A. C. Lasaga. 1993. Cooling in the late Cenozoic. *Nature* 361: 123–4.

Caldeira, K. and J. F. Kasting. 1992a. Susceptibility of the early Earth to irreversible glaciation caused by carbon dioxide clouds. *Nature* 359: 226–28.

Caldeira, K. and J. F. Kasting. 1992b. The life span of the biosphere revisited. *Nature* 360: 721–3.

Caldeira, K. and M. Searle. 1997. Cenozoic climate and tectonics of the Himalayas and Tibet. (preprint).

Calderwood, A. R. 1998. Sm-Nd isotopic modeling the evolution of the Earth's depleted mantle and crust: estimates of continental crust recycling and accretion rates. Geological Society of America, Abstracts with Programs, Annual Meeting, Toronto. A-207.

Campbell, S. E. 1979. Soil stabilization by a prokaryotic desert crust: Implications for Precambrian land biota. Origins Life 9: 335–48.

Carver, J. H. and I. M. Vardavas. 1994. Precambrian glaciations and the evolution of the atmosphere. Ann Geophys 12: 674–82.

Carver, J. H. and I. M. Vardavas. 1995. Atmospheric carbon dioxide and the long-term control of the Earth's climate. Ann Geophys 13: 782–90.

Casey, W. H. and G. Sposito. 1992. On the temperature dependence of mineral dissolution rates. Geochim Cosmochim Acta 56: 3825–30.

Castresana, J. and M. Saraste. 1995. Evolution of energetic metabolism: The respiration—early hypothesis. Trends Biochem Sci 20: 443–48.

Cawley, J. L., R. C. Burruss, and H. D. Holland. 1969. Chemical weathering in Central Iceland: An analog of pre-Silurian weathering. Science 165: 391–2.

Cerling, T. E. 1992. Use of carbon isotopes in paleosols as an indicator of the $p(CO_2)$ of the paleoatmosphere. Global Biogeochem Cycles 6: 307–14.

Chamberlin, T. C., 1899. An attempt to frame a working hypothesis of the cause of glacial periods on an atmospheric basis. J Geol 7: 545–84, 667–85, 751–87.

Chapman, D. J. 1992. "Origin and Divergence of Protists." In: J. W. Schopf and C. Klein, eds., The Proterozoic Biosphere, pp. 477–83. New York: Cambridge University Press.

Charlson, R. J. 1991. "Atmospheric Sulfur from Oceanic Phytoplankton versus Sulfur from Industry: Which Dominates Cloud Condensation Nuclei?" In: S. H. Schneider and P. J. Boston, eds., Scientists on Gaia, pp. 147–52. Cambridge, MA: MIT Press.

Charlson, R. J., J. E. Lovelock, M. O. Andreae, and S. G. Warren. 1987. Oceanic phytoplankton, atmospheric sulphur, cloud albedo and climate. Nature 326: 655–61.

Chyba, C. F. 1997. Life on other moons. Nature 385: 201.

Chyba, C. F., P. J. Thomas, L. Brookshaw, and C. Sagan. 1990. Cometary delivery of organic molecules to the early Earth. Science 249: 366–73.

Cloud, P. 1976. Beginnings of biospheric evolution and their biogeochemical consequences. Paleobiology 2: 351–87.

Cochran, M. F., F. April, F. H. Bormann, R. Berner, B. Bormann, B. Bowden,

M. Snyder, and D. Wang. 1996. Toward comprehensive chemical weathering budgets for the Hubbard Brook sandbox experiment, New Hampshire [Abstract GS31B-2]. *EOS Am Geophys Union* 77(Suppl): S94.

Cochran, M. F. and R. A. Berner. 1992. Initial effects of vegetation on Hawaiian basalt weathering rates [Abstract]. Geological Society of America Program, Boulder, CO, Annual Meeting 24, No. 7, p. A169.

Cochran, M. F. and R. A. Berner. 1993a. Reply to the Comments on "Weathering, plants, and the long-term carbon cycle." *Geochim Cosmochim Acta* 57: 2147–8.

Cochran, M. F. and R. A. Berner. 1993b. Enhancement of silicate weathering rates by vascular land plants: Quantifying the effect. *Chem Geol* 107: 213–5.

Cochran, M. F. and D. Schwartzman. 1995. Chemical weathering rates on lichen-colonized silicate rocks. Annual Meeting, Geological Society of America, Boulder, CO, Abstracts with Programs 27, No. 6, p. A-185.

Cody, R. D. 1976. Growth and early diagenetic changes in artificial gypsum crystals grown within bentonite muds and gels. *Geol Soc Am Bull* 87: 1163–8.

Cody, R. D. 1979. Lenticular gypsum: occurrences in nature, and experimental determinations of effects of soluble green plant material on its formation. *J Sed Petrol* 49: 1015–28.

Cody, R. D. and A. M. Cody. 1988. Gypsum nucleation and crystal morphology in analog saline terrestrial environments. *J Sed Petrol* 58: 247–55.

Cohen, J. E. and A. D. Rich. 1998. "Daisyworld with Interspecific Competition." In: A. Farina, J. Kennedy, and V. Bossu, eds., *Proceedings of the VII Congress of Ecology.* July 19–25, Firenze, Italy, p. 92. Firma Effe: Reggio Emilia.

Colin, F., G. H. Brimhall, D. Nahon, C. J. Lewis, A. Baronnet, and K. Danti. 1992. Equatorial rain forest lateritic mantles: A geomembrane filter. *Geology* 20: 523–6.

Collerson, K. D. and B. S. Kamber. 1999. Evolution of the continents and the atmosphere inferred from Th-U-Nb systematics of the depleted mantle. *Science* 283: 1519–22.

Colman, S. M. and K. L. Pierce. 1981. Weathering rinds on andesitic and basaltic stones as a Quaternary age indicator, Western United States. Geological Survey Professional Paper 1210.

Crowley, T. J. 1983. The geologic record of climatic change. *Rev Geophys Space Phys* 21: 828–77.

Crowley, T. J. and G. R. North. 1991. *Paleoclimatology.* New York: Oxford University Press.

Darwin, C. 1881. (reprinted 1948). The Formation of Vegetable Mould Through the Action of Worms with Observations on their Habits. (Reprinted as *Darwin on Humus and the Earthworm.*) London: Faber & Faber.

Das Sharma, S., D. J. Patil, R. Srinivasan, and K. Gopalan. 1994. Very high ^{18}O enrichment in Archean cherts from south India: Implications for Archean ocean temperature. *Terra Nova* 6: 385–90.

Davidson, E. H., K. J. Peterson, and R. A. Cameron. 1995. Origin of bilaterian body plans: Evolution of developmental regulatory mechanisms. *Science* 270: 1319–25.

Dawkins, R. 1982. *The Extended Phenotype*. New York: Freeman.

de Ronde, C. E. J. and T. W. Ebbesen. 1996. 3.2 b. y. of organic compound formation near sea-floor hot springs. *Geology* 24: 791–4.

Deines, P. 1992. "Mantle Carbon: Concentration, Mode of Occurrence, and Isotopic Composition." In: M. Schidlowski, S. Golubic, M. M. Kimberly, D. M. McKirdy, and P. A. Trudinger, eds., *Early Organic Evolution: Implications for Mineral and Energy Resources.*, pp. 133–146. Berlin: Springer-Verlag.

Delsemme, A. H. 1992. Cometary origin of carbon, nitrogen and water on the Earth. *Origins Life Evolution Biosphere* 21: 279–98.

Derry, L. A. and C. France-Lanord. 1995. Changing ^{87}Sr/^{86}Sr of Himalayan rivers during the Neogene: Implications for erosion history, uplift and the effect of the monsoon climate. Annual Meeting, Geological Society of America, Abstracts with Programs 27, No.6, p. A-237.

Des Marais, D. J. 1985. "Carbon Exchange Between the Mantle and Crust, and Its Effect upon the Atmosphere: Today Compared to Archean Time." In: E. T. Sunquist and W. S Broecker, eds., *The Carbon Cycle and Atmospheric CO_2: Natural Variations Archean to Present,* Geophysical Monograph 32, pp. 602–11. Washington, DC: American Geophysics Union.

Des Marais, D. J., H. Strauss, R. E. Summons, and J. M. Hayes. 1992. Carbon isotope evidence for the stepwise oxidation of the Proterozoic environment. *Nature* 359: 605–9.

Digregorio, B. E. 1997. *Mars: The Living Planet*. Berkeley, CA: Frog Ltd.

Dole, S. H. 1964. *Habitable Planets for Man*. New York: Blaisdell Publishing.

Doolittle, R. F. 1998b. Microbial genomes opened up. *Nature* 392: 339–42.

Doolittle, W. F. 1981. Is nature really motherly? *Coevolution Q* 29: 58–65.

Doolittle, W. F. 1998a. A paradigm gets shifty. *Nature* 392: 15–6.

Dove, P. M. 1995. "Kinetic and Thermodynamic Controls on Silica Reactivity in Weathering Environments." In: A. F. White and S. L. Brantley, eds., *Chemical Weathering Rates of Silicate Minerals. Reviews in Mineralogy,* Vol. 31, pp. 235–90. Washington, DC: Mineralogical Society of America.

Doyle, L., C. P. McKay, D. P. Whitmire, J. J. Matese, R. T. Reynolds, and W. L.

Davis. 1993. "Astrophysical Constraints on Exobiological Habitats." In: G. S. Shostak, ed., *Third Decennial US-USSR Conference on SETI, ASP Conference Series*, Vol. 47, pp. 199–217. San Francisco: Astronomical Society of the Pacific.

Drever, J. I. 1988. *The Geochemistry of Natural Waters*, 2nd ed. Englewood Cliffs, NJ: Prentice-Hall.

Drever, J. I. 1994. The effect of land plants on weathering rates of silicate minerals. *Geochim Cosmochim Acta* 58: 2325–32.

Drever, J. I. and D. W. Clow. 1995. "Weathering Rates in Catchments." In: A. F. White and S. L. Brantley, eds., *Chemical Weathering Rates of Silicate Minerals. Reviews in Mineralogy*, Vol. 31, pp. 463–83. Washington, DC: Mineralogical Society of America.

Drever, J. I. and D. R. Hurcomb. 1986. Neutralization of atmospheric acidity by chemical weathering in an alpine basin in the North Cascade Mountains. *Geology* 14: 221–4.

Drever, J. I., Y-H. Li, and J. B. Maynard. 1988. "Geochemical Cycles: The Continental Crust and the Oceans." In: C. B. Gregor, R. M. Garrels, F. T. Mackenzie, and J. B. Maynard, eds., *Chemical Cycles in the Evolution of the Earth*, pp. 17–53. New York: Wiley & Sons.

Drever, J. I. and G. F. Vance. 1994. "Role of Organic Acids in Mineral Weathering Processes." In: E. D. Pittman and M. D. Lewan, eds., *Organic Acids in Geological Processes*, pp. 138–161. Berlin: Springer-Verlag.

Drever, J. I. and J. Zobrist. 1992. Chemical weathering of silicate rocks as a function of elevation in the southern Swiss Alps. *Geochim Cosmochim Acta* 56: 3209–16.

Dyer, B. D. and R. A. Obar. 1994. *Tracing the History of Eukaryotic Cells: The Enigmatic Smile*. New York: Columbia University Press.

Easterbrook, G. 1995. *A Moment on Earth: The Coming Age of Environmental Optimism*. New York: Viking.

Ebeling, W. 1985, Thermodynamics of self-organization and evolution. *Biomed Biochim Acta* 44: 831–8.

Eckhardt, F. E. W. 1985. "Solubilization, Transport, and Deposition of Mineral Cations by Microorganisms—Efficient Rock Weathering Agents." In: J. I. Drever, ed., *The Chemistry of Weathering*, pp. 161–73, Dordrecht: D. Reidel Publishers.

Edmond, J. M. 1992. Himalayan tectonics, weathering processes, and the strontium isotope record in marine limestones. *Science* 258: 1594–1597.

Edmond, J. M. and Y. Huh. 1997. A critique of geochemical models of the Cenozoic/Phanerozoic evolution of atmospheric pCO_2. *Rev Geophys* (submitted for publication).

Edmond, J. M., M. R. Palmer, C. I. Measures, B. Grant, and R. F. Stallard. 1995. The fluvial geochemistry and denudation rate of the Guayana Shield in Venezuela, Colombia, and Brazil. *Geochim Cosmochim Acta* 59: 3301–25.

Eggleston, C. M., M. F. Hochella Jr., and G. A. Parks. 1989. Sample preparation and aging effects on dissolution rate and surface composition of diopside. *Geochim Cosmochim Acta* 53: 797–804.

Elderfield, H. and A. Schultz. 1996. Mid-ocean ridge hydrothermal fluxes and the chemical composition of the ocean. *Annu Rev Earth Planet Sci* 24: 191–224.

Eldridge, D. J. and R. S. B. Greene. 1994. Assessment of sediment yield by splash erosion on a semi-arid soil with varying cryptogam cover. *J Arid Environ* 26: 221–32.

Elick, J. M., S. G. Driese, and C. I. Mora. 1998. Very large plant and root traces from the Early to Middle Devonian: Implications for early terrestrial ecosystems and atmospheric p(CO_2). *Geology* 26: 143–6.

Engels, F. 1940. *Dialectics of Nature*. New York: International Publishers.

Epstein, S. and J. Karhu. 1990. A reply to E. C. Perry's comments. *Geochim Cosmochim Acta* 54: 1181–4.

Essex, C. 1984, Radiation and the irreversible thermodynamics of climate. *J Atmospheric Sci* 41: 1985–91.

Evans, D. A., N.J. Beukes, and J. L. Kirschvink. 1997. Low-latitude glaciation in the Paleaoproterozoic era. *Nature* 386: 262–6.

Fahey, B. D. 1983. Frost action and hydration as rock weathering mechanisms on schist: A laboratory study. *Earth Surface Processes Landforms* 8: 535–45.

Fairbairn, H. W., P. M. Hurley, K. D. Card, and C. J. Knight. 1969. Correlation of radiometric ages of Nipissing diabase and Huronian metasediments with Proterozoic orogenic events in Ontario. *Can J Earth Sci* 6: 489–97.

Fedo, C. M., K. A. Eriksson, and E. J. Krogstad. 1996. Geochemistry of shales from the Archean (3.0 Ga) Buhwa Greenstone Belt, Zimbabwe: Implications for provenance and source-area weathering. *Geochim Cosmochim Acta* 60: 1751–63.

Fedonkin, M. A., E. L. Yochelson, and R. J. Horodyski. 1994. Ancient metazoa. *Natl Geogr Res Explor* 10: 200–23.

Feinberg, G. and R. Shapiro. 1980. *Life Beyond Earth*. New York: William Morrow.

Fell, N. and P. Liss. 1993. Can algae cool the planet? *New Scientist* (8/21/93): 34–8.

Fontana, W. and L. W. Buss. 1994. What would be conserved if "the tape were played twice"? *Proc National Acad Sci USA* 91: 757–61.

Forget, F, and R. T. Pierrehumbert. 1997. Warming early Mars with carbon dioxide clouds that scatter infrared radiation. *Science* 278: 1273–6.

Forterre, P. 1995a. The reverse gyrase of hyperthermophilic archaeobacteria: Origin of life and thermophily. *Microbiol SEM* 11: 225–32.

Forterre, P. 1995b. Thermoreduction, a hypothesis for the origin of prokaryotes. *C R Acad Sci Paris* (Life Sci) 318: 415–22.

Forterre, P. 1995c. Looking for the most "primitive" organism(s) on Earth today: The state of the art. *Planet Space Sci* 43: 167–77.

Forterre, P., F. Confalonieri, F. Charbonnier, and M. Duguet. 1995. Speculations on the origin of life and thermophily: Review of available information on reverse gyrase suggests that hyperthermophilic procaryotes are not so primitive. *Origins Life Evolution Biosphere* 25: 235–49.

Frakes, L. A., J. E. Francis, and J. I. Syktus. 1992. *Climate Modes of the Phanerozoic.* New York: Cambridge University Press.

Francois, L. M. and J. C. G. Walker. 1992, Modelling the Phanerozoic carbon cycle and climate: Constraints from the $^{87}Sr/^{86}Sr$ isotopic ratio of seawater. *Am J Sci* 292: 81–135.

Fry, E. J. 1924, A suggested explanation of the mechanical action of lithophytic lichens on rocks (shale). *Ann Botany* (London) 38: 175–96.

Fry, E. J. 1927. The mechanical action of crustaceous lichens on substrata of shale, schist, gneiss, limestone and obsidian. *Ann Botany* (London) 41: 437–60.

Gaillardet, J., B. Dupre, and C. J. Allegre. 1995. A global geochemical mass budget applied to the Congo Basin rivers: Erosion rates and continental crust composition. *Geochim Cosmochim Acta* 59: 3469–85.

Gaillardet, J., P. Louvat, B. Dupre, and C. J. Allegre. 1997. "Coupling Modern Chemical Weathering Rates of Silicates Using Rivers." LPI Contribution No. 921. In: *Seventh Annual V. M. Goldschmidt Conference,* p. 77. Houston, TX: Lunar and Planetary Institute.

Galimov, E. M., A. A. Migdisov, and A. B. Ronov. 1975. Variation in the isotopic composition of carbonate and organic carbon in sedimentary rocks during Earth's history. *Geokhimiya* 11: 323–42; *Geochem Int* 12: 1–19.

Galtier, N., N. Tourasse, and M. Gouy. 1999. A nonhyperthermophilic common ancestor to extant life forms. *Science* 283: 220–1.

Garrels, R. M. 1987. A model for the deposition of the microbanded Precambrian iron formations. *Am J Sci* 287: 81–106.

Gat, J. R. 1996. Oxygen and hydrogen isotopes in the hydrologic cycle. *Ann Rev Earth Planet Sci* 24: 225–62.

Geiger, R., R. H. Aron, and P. Todhunter. 1995. *The Climate Near the Ground,* 5th ed. Wiesbaden, Germany: Vieweg.

Gerald, J-C., and V. Dols. 1990. The warm Cretaceous climate: Role of the long-term carbon cycle. *Geophys Res Lett* 17: 1561–4.

Gerlach, T. 1991. Etna's greenhouse pump. *Nature* 351: 352–3.

Ghilarov, A. M. 1995. Vernadsky's biophere concept: An historical perspective. *Q Rev Biol* 70: 193–203.

Gislason, S. R., S. Arnorsson, and H. Armannsson. 1996. Chemical weathering of basalt in southwest Iceland: Effects of runoff, age of rocks and vegetative/glacial cover. *Am J Sci* 296: 837–907.

Gislason, S. R. and H. P. Eugster. 1987. Meteoric water-basalt interactions: II. A field study in N.E. Iceland. *Geochim Cosmochim Acta* 51: 2841–56.

Gobran, G. R., S. Clegg, and F. Courchesne. 1998. Rhizospheric processes influencing the biogeochemistry of forest ecosystems. *Biogeochemistry* 42: 107–20.

Godderis, Y. and L. M. Francois. 1995. The Cenozoic evolution of the strontium and carbon cycles: Relative importance of continental erosion and mantle exchanges. *Chem Geol* 126: 169–90.

Gogarten-Boekels, M., E. Hilario, and J. P. Gogarten. 1995. The effects of heavy meteorite bombardment on the early evolution—the emergence of the three domains of life. *Origins Life Evolution Biosphere* 25: 251–64.

Gold, T. 1992. The deep, hot biosphere. *Proc Natl Acad Sci U S A* 89: 6045–9.

Gorshkov, G. and A. Yakushova. 1972. *Physical Geology.* Moscow: Mir.

Gould, S. J. 1985. *The Flamingo's Smile.* New York: Norton.

Gould, S. J. 1989. *Wonderful Life: The Burgess Shale and the Nature of History.* New York: Norton.

Gould, S. J. 1996. *Full House.* New York: Harmony Books.

Graedel, T. E., I-J. Sackmann, and A. I. Boothroyd. 1991, Early solar mass loss: A potential solution to the weak sun paradox. *Geophys Res Lett* 18: 1881–4.

Grandstaff, D. E. 1986. "The Dissolution Rate of Forsteritic Olivine from Hawaiian Beach Sand." In: S. Coleman and D. Dethier, eds., *Rates of Chemical Weathering of Rocks and Minerals,* pp. 41–59. New York: Academic Press.

Gray, M. W., G. Burger and B. Franz Lang. 1999. Mitochondrial evolution. *Science* 283: 1476–81.

Grinevald, J. 1988. "Sketch for a History of the Biosphere." In: P. Bunyard and E. Goldsmith, eds., *Gaia, the Thesis, the Mechanisms and the Implications,* pp. 1–34, Wadebridge, Great Britain: Quintrell.

Grotzinger, J. P. 1994. "Trends in Precambrian Carbonate Sediments and Their Implication for Understanding Evolution." In: S. Bengtson, ed., *Early Life on*

Earth. Nobel Symposium No. 84, pp. 245–58, New York: Columbia University Press.

Grotzinger, J. P. and J. F. Kasting. 1993. New constraints on Precambrian ocean composition. *J Geol* 101: 235–43.

Gunatilaka, A. 1990. Anhydrite diagenesis in a vegetated sabkha, Al-Khiran, Kuwait, Arabian Gulf. *Sed Geol* 69: 95–116.

Gutzmer, J. and N.J. Beukes. 1998. Earliest laterites and possible evidence for terrestrial vegetation in the Early Proterozoic. *Geology* 26: 263–6.

Gwiazda, R. H. and W. S. Broecker. 1994. The separate and combined effects of temperature, soil pCO_2, and organic acidity on silicate weathering in the soil environment: Formulation of an model and results. *Global Biogeochem Cycles* 8: 141–55.

Haldane, J. B. S. 1985. *On Being the Right Size and Other Essays*. Oxford, England: Oxford University Press.

Halliday, A. N., D-C. Lee, J. N. Christensen, M. Rehkamper, W. Yi, and X. Luo. 1997. "Plasmas, the Early Solar System, and Climate Dynamics." LPI Contribution No. 921. In: *Seventh Annual V. M. Goldschmidt Conference*, p. 85. Houston, TX: Lunar and Planetary Institute.

Hamilton, W. D. and T. M. Lenton. 1998. Spora and Gaia: how microbes fly with their clouds. *Ethol Ecol Evolution* 10: 1–16.

Han, T-M., and B. Runnegar. 1992. Megascopic eukaryotic algae from the 2.1-billion-year-old Negaunee iron-formation, Michigan. *Science* 257: 232–5.

Hardie, L. A. 1996. Secular variation in seawater chemistry: An explanation for the coupled secular variation in the mineralogies of marine limestones and potash evaporites over the past 600 m. y. *Geology* 24: 279–83.

Harland, W. B. 1983. "The Proterozoic Glacial Record." In: L. G. Medaris, C. W. Byers, D. M. Mickelson, and W. C. Shanks, eds., *Proterozoic Geology: Selected Papers from an International Proterozoic Symposium*, GSA Memoir 161. pp. 279–88. Boulder, CO: Geological Society of America.

Harris, N. 1995. Significance of weathering Himalayan metasedimentary rocks and leucogranites for the Sr isotope evolution of seawater during the early Miocene. *Geology* 23: 795–8.

Hartenstein, R. 1986. "Earthworm Biotechnology and Global Biogeochemistry." In: A. Macfadyen and E. D. Ford, eds., *Advances in Ecological Research 15*, pp. 379–409. London: Academic Press.

Hauri, E. H., N. Shimizu, J. J. Dieu, and S. R. Hart. 1993. Evidence for hotspot-related carbonatite metasomatism in the oceanic upper mantle. *Nature* 365: 221–7.

Hawksworth, D. L. 1988. "The Fungal Partner." In: M. Galun, ed., *CRC Handbook of Lichenology*, Vol. I, pp. 35–8. Boca Raton, FL: CRC Press.

Hayashi, K-I., H. Fujisawa, H. D. Holland, and H. Ohmoto. 1997. Geochemistry of 1.9 Ga sedimentary rocks from northeastern Labrador, Canada. *Geochim Cosmochim Acta* 61: 4115–37.

Hayes, J. M. 1994. "Global Methanotrophy at the Archean-Proterozoic Transition." Nobel Symposium No. 84. In: S. Bengtson, ed., *Early Life on Earth*, pp. 220–36. New York: Columbia University Press.

Hedges, J. 1992. Global biogeochemical cycles: Progress and problems. *Marine Chem* 39: 67–93.

Heins, W. A. 1995. The use of mineral interfaces in sand-sized rock fragments to infer ancient climate. *Geol Soc Am Bull* 107: 113–25.

Henderson-Sellers, A. and K. McGuffie. 1987. *A Climate Primer.* Chichester, England: Wiley.

Henderson-Sellers, B., S. M. P. Benbow, and A. Henderson-Sellers. 1991. Earth-the water planet: A lucky coincidence? In: S. H. Schneider and P. J. Boston, eds., *Scientists on Gaia*, pp. 80–9. Cambridge, MA: MIT Press.

Hiebert, F. K. and P. C. Bennett. 1992. Microbial control of silicate weathering in organic-rich ground water. *Science* 258: 278–81.

Hinrichs, K-U., J. M. Hayes, S. P. Sylva, P. G. Brewer, and E. F. DeLong. 1999. Methane-consuming archaebacteria in marine sediments. *Nature* 398: 802–5.

Ho, M-W. 1993. *The Rainbow and the Worm.* Singapore: World Scientific.

Ho, M-W. 1995. Bioenergetics and the coherence of organisms. *Neural Network World* (April): 733–50.

Hochella, M. F. Jr., and J. F. Banfield. 1995. "Chemical Weathering of Silicates in Nature: A Microscopic Perspective with Theoretical Considerations." In: A. F. White and S. L. Brantley, eds., *Chemical Weathering Rates of Silicate Minerals. Reviews in Mineralogy*, Vol. 31, pp. 353–406. Washington, DC: Mineralogical Society of America.

Hoffert, M. I. and C. Covey. 1992. Deriving global climate sensitivity from palaeoclimate reconstructions. *Nature* 360: 573–6.

Hofmann, A. W. 1997. Early evolution of the continents. *Science* 275: 498–9.

Hofmann, H. J. 1994. "Proterozoic Carbonaceous Compressions ('Metaphytes' and 'Worms')." In: S. Bengtson, ed., *Early Life on Earth*, Nobel Symposium No. 84. pp. 342–57. New York: Columbia University Press.

Holland, H. D. 1978. *The Chemistry of the Atmosphere and Oceans.* New York: Wiley & Sons.

Holland, H. D. 1994. "Early Proterozoic Atmospheric Change." In: S. Bengtson, ed., *Early Life on Earth*, Nobel Symposium No. 84. pp. 237–44. New York: Columbia University Press.

Holland, H. D. and R. Rye. 1997. Evidence in pre-2.2 Ga paleosols for the early evolution of atmospheric oxygen and terrestrial biota: Comment. *Geology* 25: 857–8.

Holldobler, B. and E. O. Wilson. 1990. *The Ants*. Cambridge, MA: Belknap Press of Harvard University Press.

Holmden, C. and K. Muehlenbachs. 1993. The $^{18}O/^{16}O$ ratio of 2-billion-year-old seawater inferred from ancient oceanic crust. *Science* 259: 1733–6.

Holmen, K. 1992. "The Global Carbon Cycle." In: S. S. Butcher, R. J. Charlson, G. H. Orians, and G. V. Wolfe, eds., *Global Biogeochemical Cycles*, pp. 239–62. London: Academic Press.

Holmes, A. 1978. *Principles of Physical Geology*, 3rd ed. Walton-on-Thames, England: Thomas Nelson & Sons.

Holser, W. T., M. Shidlowski, F. T. Mackenzie, and J. B. Maynard. 1988. "Geochemical Cycles of Carbon and Sulfur." In: C. B. Gregor, R. M. Garrels, F. T. Mackenzie, and J. B. Maynard, eds., *Chemical Cycles in the Evolution of the Earth*, pp. 105–73. New York: Wiley & Sons.

Horodyski, R. J. and L. P. Knauth. 1994. Life on land in the Precambrian. *Science* 263: 494–98.

Hoyle, F. 1972. The history of the Earth. *Q J R Astron Soc* 13: 328–45.

Huh, Y. and J. M. Edmond. 1997. "Chemical Weathering Yields and Strontium Isotope Systematics from Major Siberian Rivers." LPI Contribution No. 921. *Seventh Annual V. M. Goldschmidt Conference*, p. 100. Houston, TX: Lunar and Planetary Institute.

Hutchinson, G. E. 1954. "The Biochemistry of the Terrestrial Atmosphere." In: G. P. Kuiper, ed., *The Earth as a Planet*, pp. 371–433. Chicago: University of Chicago Press.

Jackson, T. A. 1968. The role of pioneer lichens in the chemical weathering of recent volcanic rocks on the island of Hawaii [Ph.D. dissertation]. University of Missouri, Columbia.

Jackson, T. A. 1993. Comment on "Weathering, plants, and the long-term carbon cycle" by Robert A. Berner. *Geochim Cosmochim Acta* 57: 2141–4.

Jackson, T. A. and W. D. Keller. 1970a. A comparative study of the role of lichens and inorganic processes in the chemical weathering of recent Hawaiian lava flows. *Am J Sci* 269: 446–66.

Jackson, T. A. and W. D. Keller. 1970b. Evidence for biogenic synthesis of an unusual ferric oxide mineral during alteration of basalt by a tropical lichen. *Nature* 227: 522–3.

Jacobsen, S. B. 1988. Isotopic and chemical constraints on mantle-crust evolution. *Geochim Cosmochim Acta* 52: 1341–50.

Jacobsen, S. B. and M. R. Pimentel-Klose. 1988a. Nd isotopic variations in Precambrian banded iron formations. *Geophys Res Lett* 15: 393–6.

Jacobsen, S. B. and M. R. Pimentel-Klose. 1988b. A Nd isotopic study of the Hamersley and Michipicoten banded iron formations: The source of REE and Fe in Archean oceans. *Earth Planet Sci Lett* 87: 29–44.

Jean-Baptiste, P., J. L. Charlou, and M. Stievenard. 1997. Oxygen isotope study of mid-ocean ridge hydrothermal fluids: Implication for the oxygen-18 budget of the ocean. *Geochim Cosmochim Acta* 61: 2669–77.

Jenkins, G. S., H. G. Marshall, and W. R. Kuhn. 1993. Precambrian climate: The effects of land area and the Earth's rotation rate. *J Geophys Res* 98: 8785–91.

Johansson, A. K. 1993. Grasslands, silicate weathering and diatoms: Cause and effect? Geological Society of America, Abstracts with Programs, MidCentral Section. No. 27621.

Johansson, A. K. 1995. Plant mediated weathering: The driving force behind increased terrestrial and marine productivity, and Neogene climate change? Geological Society of America, Abstracts with Programs, Northeastern Section. Vol.27, No. 1, p. 58.

Johansson, A. K. 1996. A comparative study of biologically mediated silicate weathering [M.S. dissertation]. Columbia University, New York.

Johansson, A. K., P. N. Froelich, and C. H. Langmuir. 1997. Biologically-mediated silica polymerization in laboratory phytotron experiments: Implications for mineral weathering [Abstract OS41B-08]. *EOS Trans Am Geophys Union* 78(Suppl): F369.

Johnston, C. G. and J. R. Vestal. 1989. Distribution of inorganic species in two Antarctic cryptoendolithic microbial communities. *Geomicrobiol J* 7: 137–53.

Johnston, C. G. and J. R. Vestal. 1993, Biogeochemistry of oxalate in the Antarctic cryptoendolithic lichen-dominated community. *Microbial Ecol* 25: 305–19.

Jones, D. 1988. "Lichens and Pedogenesis." In: M. Galun, ed., *CRC Handbook of Lichenology*, Vol. III, pp. 109–24. Boca Raton, FL: CRC Press.

Jones, D. and M. J. Wilson. 1985. Chemical activity of lichens on mineral surfaces— A review. *Int Biodeterioration* 21: 99–104.

Jones, D., M. J. Wilson, and J. M. Tait. 1980. Weathering of basalt by Pertusaria corallina. *Lichenologist* 12: 277–89.

Jones, D. L. and L. P. Knauth. 1979. Oxygen isotopic and petrographic evidence relevant to the origin of the Arkansas novaculite. *J Sed Petrol* 49: 581–98.

Jongmans, A. G., N. van Breemen, U. Lundstrom, P. A. W. van Hees, R. D. Finlay, M. Srinivasan, T. Unestam, R. Giesler, P-A. Melkerud, and M. Olsson. 1997. Rock-eating fungi. *Nature* 389: 682–3.

Kamshilov, M. M. 1976. *Evolution of the Biosphere.* Moscow: Mir.

Karhu J. and S. Epstein. 1986. The implication of the oxygen isotope records in coexisting cherts and phosphates. *Geochim Cosmochim Acta* 50: 1745–56.

Kasting, J. F. 1987. Theoretical constraints on oxygen and carbon dioxide concentrations in the Precambrian atmosphere. *Precambr Res* 34: 205–29.

Kasting, J. F. 1989. Long-term stability of the Earth's climate. *Palaeogeogr Palaeoclimatol Palaeoecol* (Global Planetary Change Section) 75: 83–95.

Kasting, J. F. 1992. "Proterozoic Climates: The Effect of Changing Atmospheric Carbon Dioxide Concentrations." In: J. W. Schopf and C. Klein, eds., *The Proterozoic Biosphere,* pp. 165–8. New York: Cambridge University Press.

Kasting, J. F. 1996. "Habitable Zones Around Stars: An Update." In: L. R. Doyle, ed., *Circumstellar Habitable Zones,* pp. 17–28. Menlo Park, CA: Travis House Publishers.

Kasting, J. F. 1997. Warming early Earth and Mars. *Science* 276: 1213–5.

Kasting, J. F. and T. P. Ackerman. 1986. Climatic consequences of very high CO_2 levels in Earth's early atmosphere. *Science* 234: 1383–5.

Kasting, J. F., D. H. Eggler, and S. P. Raeburn. 1993. Mantle redox evolution and the oxidation state of the Archean atmosphere. *J Geol* 101: 245–57.

Kasting, J. F. and D. H. Grinspoon. 1990. "The Faint Young Sun Problem." In: C. P. Sonnett, M. S. Giampapa, and M. S. Matthews, eds., *The Sun in Time,* pp. 447–62. Tucson, AZ: University of Arizona Press.

Kasting, J. F. and O. B. Toon. 1989. "Climate Evolution on the Terrestrial Planets." In: S. K. Atreya, J. B. Pollack, and M. S. Matthews, eds., *Origin and Evolution of Planet and Satellite Atmospheres,* pp. 423–49. Tucson, AZ: University Arizona Press.

Kasting, J. F., O. B. Toon, and J. B. Pollack. 1988. How climate evolved on the terrestrial planets. *Sci Am* 258: 90–7.

Kasting, J. F., D. P. Whitmire, and R. T. Reynolds. 1993. Habitable zones around main sequence stars. *Icarus* 101: 108–28.

Kasting, J. F., K. J. Zahnle, J. P. Pinto, and A. T. Young. 1989. Sulfur, ultraviolet radiation, and the early evolution of life. *Origins Life Evolution Biosphere* 19: 95–108.

Kasting, J. F., K. J. Zahnle, and J. C. G. Walker. 1983. Photochemistry of methane in the Earth's early atmosphere. *Precambrian Res* 20: 121–48.

Kauffman, S. A. 1993. *The Origins of Order.* New York: Oxford University Press.

Kauffman, S. A. 1995. *At Home in the Universe: The Search for Laws of Self-Organization.* New York: Oxford University Press.

Keller, C. K. and B. D. Wood. 1993. Possiblity of chemical weathering before the advent of vascular land plants. *Nature* 364: 223–5.

Kelly, E. F., O. A. Chadwick, and T. E. Hilinski. 1998. The effect of plants on mineral weathering. *Biogeochemistry* 42: 21–53.

Kenny, R. and L. P. Knauth. 1992. Continental paleoclimates from δD and $\delta^{18}O$ of secondary silica in paleokarst chert lags. *Geology* 20: 219–22.

Kieft, T. L. 1988. Ice nucleation activity in lichens. *Appl Environ Microbiol* 54: 1678–81.

Kirkley, M. B., J. J. Gurney, M. L. Otter, S. J. Hill, and L. R. Daniels. 1991. The application of C isotope measuresments to the identification of sources of C in diamonds: A review. *Appl Geochem* 6: 477–94.

Kirchner, J. W. 1989. The Gaia hypothesis: can it be tested? *Rev Geophys* 27: 223–35.

Klein, C. and O. P. Bricker. 1977. Some aspects of the sedimentary and diagenetic environment of Proterozoic banded iron-formations. *Econ Geol* 72: 1457–70.

Klinger, L. F. 1991. "Peatland Formation and Ice Ages: A Possible Gaian Mechanism Related to Community Succession." In: S. H. Schneider and P. J. Boston, eds., *Scientists on Gaia,* pp. 247–55, Cambridge, MA: MIT Press.

Klinger, L. F. and D. J. Erickson III. 1997. Geophysiological coupling of marine and terrestrial ecosystems. *J Geophys Res* 102: 25,359–70.

Knauth, L. P. 1992. Origin and diagensis of cherts: "An Isotopic Perspective." In: N. Clauer and S. Chaudhuri, eds., *Isotopic Signatures and Sedimentary Records, Lecture Notes in Earth Sciences #43,* pp. 123–52. New York: Springer-Verlag.

Knauth, L. P. 1998. Salinity history of the Earth's early ocean. *Nature* 395: 554–5.

Knauth, L. P. and D. R. Lowe. 1978. Oxygen isotope geochemistry of cherts from the Onverwacht Group (3.4 billion years), Transvaal, South Africa, with implications for secular variations in the isotopic composition of cherts. *Earth Planet Sci Lett* 41: 209–22.

Knauth, L. P. and P. L. Clemens. 1995. Climatic history of the Earth based on isotopic analyses of cherts. Annual Meeting, Geological Society of America, Abstracts with Programs 27, No. 6, p. A-205.

Knauth, L. P. and S. Epstein. 1976. Hydrogen and oxygen isotope ratios in nodular and bedded cherts. *Geochim Cosmochim Acta* 40: 1095–108.

Knoll, A. H. 1990. "Precambrian Evolution of Prokaryotes and Protists." In: D. E. G. Briggs and P. R. Crowther, eds., *Palaeobiology. A Synthesis,* pp. 9–16. Oxford, England: Blackwell Scientific.

Knoll, A. H. 1992. The early evolution of eukaryotes: A geological perspective. *Science* 256: 622–27.

Knoll, A. H. and T. Bauld. 1989. The evolution of ecological tolerance in prokaryotes. *Trans R Soc Edinb Earth Sci* 80: 209–23.

Koeberl, C., V. L. Masaitis, G. I. Shafranovsky, I. Gilmour, F. Langenhorst, and M. Schrauder. 1997. Diamonds from the Popigai impact structure, Russia. *Geology* 11: 967–70.

Konhauser, K. O. and F. G. Ferris. 1996. Diversity of iron and silica precipitation by microbial mats in hydrothermal waters, Iceland: Implications for Precambrian iron formations. *Geology* 24: 323–6.

Kossinna, E. 1921. Die Tiefen des Weltmeeres. Berlin Univ. Instit. f. Meereskunde, Veröff., N.F.A. Geogr.-naturwiss. Reihe, Heft 9.

Kristjansson, J. K. and K. O. Stetter. 1992. "Thermophilic Bacteria." In: J. K. Kristjansson, ed., *Thermophilic Bacteria,* pp. 1–18. Boca Baton, FL: CRC Press.

Krumbein, W. E. and B. D. Dyer. 1985. "This Planet is Alive—Weathering and Biology, A Multi-facetted Problem." In: J. I. Drever, ed., *The Chemistry of Weathering,* pp. 143–60. Dordrecht: D. Reidel Publishers.

Krumbein, W. E. and K. Jens. 1981. Biogenic rock varnishes of the Negev Desert (Israel) an ecological study of iron and manganese transformation by cyanobacteria and fungi. *Oecologia* 50: 25–38.

Krumbein, W. E. and A. V. Lapo. 1996. "Vernadsky's Biosphere as a Basis of Geophysiology." In: P. Bunyard, ed., *Gaia in Action,* pp. 115–34. Edinburgh, Scotland: Floris Books.

Krumbein, W. E. and H-J. Schellnhuber. 1990. "Geophysiology of Carbonates as a Function of Bioplanets." In: V. I. Hekkot, S. Kempe, W. Michaelis and A. Spitzy, eds., *Facets of Modern Biogeochemistry,* pp. 5–22. New York: Springer-Verlag.

Krumbein, W. E. and H-J. Schellnhuber. 1992. Geophysiology of mineral deposits—a model for a biological driving force of global changes through Earth history. *Terra Nova* 4: 351–62.

Kuhn, W. R., J. C. G. Walker, and H. G. Marshall. 1989. The effect on Earth's surface temperature from variations in rotation rate, continent formation, solar luminosity and carbon dioxide. *J Geophys Res* 94: 11,129–36.

Kump, L. R. 1996. The physiology of the planet. *Nature* 381: 111–2.

Kump, L. R. and E. J. Barron. 1988. Geologic and geographic effects on paleoclimate. *Geol Soc Am* 20: A257.

Kump, L. R. and J. E. Lovelock. 1995. "The Geophysiology of Climate." In:
A. Henderson-Sellers, ed., *Future Climates of the World: A Modeling Perspective*,
pp. 537–53. Amsterdam: Elsevier.

Lal, R. 1990. "Soil Erosion and Land Degradation: The Global Risks." In: R. Lal
and B. A. Stewart, eds., *Advances in Soil Science*. Vol. 11, Soil Degradation, pp.
129–72. New York: Springer-Verlag.

Lamprecht, I. and A. I. Zotin, eds. 1978. *Thermodynamics of Biological Processes*. Berlin:
Walter de Gruyter.

Lapo, A. V. 1982. *Traces of Bygone Biospheres*. Moscow: Mir.

Large, P. J. 1983. *Methylotrophy and Methanogenesis*. Washington, DC: American Society for Microbiology.

Lasaga, A. C. and R. A. Berner. 1998. Fundamental aspects of quantitative models
for geochemical cycles. *Chem Geol* 145: 161–75.

Lasaga, A. C., J. M. Soler, J. Ganor, T. E. Burch, and K. L. Nagy. 1994. Chemical
weathering rate laws and global geochemical cycles. *Geochim Cosmochim Acta* 58:
2361–86.

Lenton, T. M. 1998. Gaia and natural selection. *Nature* 394: 439–47.

Lesins, G. B. 1990. On the relationship between radiative entropy and temperature
distributions. *J Atmospher Sci* 47: 795–803.

Lesins, G. B. 1991. "Radiative Entropy as a Measure of Complexity." In: S. H.
Schneider and P. Boston, eds., *Scientists on Gaia*, pp. 121–7. Cambridge, MA:
MIT Press.

Leung, I. S., W. Guo, I. Friedman, and J. Gleason. 1990. Natural occurrence of silicon carbide in a diamondiferous kimberlite from Fuxian. *Nature* 346: 352–4, 874.

Levins, R. and R. Lewontin. 1985. *The Dialectical Biologist*. Cambridge, MA: Harvard University Press.

Likens, G. E. and F. H. Bormann. 1995. *Biogeochemistry of a Forested Ecosystem*, 2nd
ed. New York: Springer-Verlag.

Louvat, P. 1997. Étude géochimique de l'érosion fluvial d'îles volcaniques à l'aide
des bilans d'éléments majeurs et traces [Ph.D. thesis]. Institute Physique du
Globe, University of Paris.

Lovelock, J. 1988. *The Ages of Gaia*. New York: W. W. Norton.

Lovelock, J. E. 1979. *Gaia*. New York: Oxford University Press.

Lovelock, J. E. 1987. "Geophysiology: A New Look at Earth Sciences." In: R. E. Dickinson, ed., *The Geophysiology of Amazonia*, pp. 11–23. New York: Wiley & Sons.

Lovelock, J. E. 1989. Geophysiology, the science of Gaia. *Rev Geophys* 27: 215–22.

Lovelock, J. E. 1990. Hands up for the Gaia hypothesis. *Nature* 344: 100–2.

Lovelock, J. E. 1991. *Healing Gaia*. New York: Harmony Books.

Lovelock, J. E. and L. R. Kump. 1994. Failure of climate regulation in a geophysiological model. *Nature* 369: 732–4.

Lovelock, J. E. and L. Margulis. 1974a. Homeostatic tendencies of the Earth's atmosphere. *Origins Life* 5: 93–103.

Lovelock, J. E. and Margulis, L. 1974b. Atmospheric homeostasis by and for the biosphere: the Gaia hypothesis. *Tellus* 26: 2–9.

Lovelock, J. E. and A. J. Watson. 1982. The regulation of carbon dioxide and climate: Gaia or geochemistry. *Planet Space Sci* 30: 795–802.

Lovelock, J. E. and M. Whitfield. 1982. Life span of the biosphere. *Nature* 296: 561–63.

Lowe, D. R. 1994. "Early Environments: Constraints and Opportunities for Early Evolution." In: S. Bengtson, ed., *Early Life on Earth*, Nobel Symposium No. 84. pp. 24–35. New York: Columbia University Press.

Lowe, D. R. and L. P. Knauth. 1977. Sedimentology of the Onverwacht Group (3.4 billion years), Transvaal, South Africa, and its bearing on the characteristics and evolution of the early earth. *J Geol* 85: 699–723.

Mackenzie, F. T. 1990. Distribution, chemistry, mineralogy and cycling of the Phanerozoic carbonate mass. V. M. Goldschmidt Conference, Program and Abstracts, May 2–4, Baltimore, MD, The Geochemical Society, p. 63.

Mackenzie, F. T. and L. R. Kump. 1995. Reverse weathering, clay mineral formation, and oceanic element cycles. *Science* 270: 586–7.

Maher, K. A. and D. J. Stevenson. 1988. Impact frustration of the origin of life. *Nature* 331: 612–4.

Manabe, S. and R. J. Stouffer. 1980. Sensitivity of a global climate model to an increase of CO_2 concentration in the atmosphere. *J Geophys Res* 85: 5529–54.

Manabe, S. and R. T. Wetherald. 1980. On the distribution of climate change resulting from an increase in CO_2-content of the atmosphere. *J Atmospher Sci* 37: 99–118.

Margulis, L. 1993. *Symbiosis in Cell Evolution*, 2nd ed. New York: W. H. Freeman.

Margulis, L. 1996. Archael-eubacterial mergers in the origin of Eukarya: Phyloge-netic classification of life. *Proc Natl Acad Sci U S A* 93: 1071–6.

Margulis, L. and R. Fester, eds. 1991. *Symbiosis as a Source of Evolutionary Innovation: Speciation and Morphogenesis.* Cambridge, MA: MIT Press.

Margulis, L. and J. E. Lovelock. 1974. Biological modulation of the earth's atmo-sphere. *Icarus* 21: 471–89.

Margulis, L. and D. Sagan. 1995. *What is Life?* New York: Simon & Schuster.

Margulis, L., J. C. G. Walker, and M. Rambler. 1976. Reassessment of roles of oxy-gen and ultraviolet light in Precambrian evolution. *Nature* 264: 620–4.

Mariotti, J. M., A. Leger, B. Mennesson, and M. Ollivier. 1997. "Detection and Characterization of Earth-like Planets." In: C. B. Cosmovici, S. Bowyer, and D. Werthimer, eds., *Astronomical and Biochemical Origins and the Search for Life in the Universe,* pp. 299–311. Bologna, Italy: Editrice Compositori.

Markos, A. 1995. The ontogeny of Gaia: The role of microorganisms in planetary information network. *J Theoret Biol* 176: 175–80.

Marmo, J. S. 1992. "The Lower Proterozoic Hokkalampi Paleosol in North Karelia, Eastern Finland." In: M. Schidlowski, S. Golubic, M. M. Kimberly, D. M. McKirdy, and P. A. Trudinger, eds., *Early Organic Evolution: Implications for Mineral and Energy Resources,* pp. 41–66. Berlin: Springer-Verlag.

Marshall, H. G., J. C. G. Walker, and W. R. Kuhn. 1988. Long-term climate change and the geochemical cycle of carbon. *J Geophys Res* 93: 791–801.

Marshall, J. R. and V. R. Overbeck. 1992. Textures of impact deposits and the origin of tillites [Abstract P31B-4]. *EOS Trans Am Geophys Union* 73(Suppl): 324.

Martin, W. and M. Muller. 1998. The hydrogen hypothesis for the first eukaryote. *Nature* 392: 37–41.

Marty, B. 1989. On mantle carbon flux calculated. *EOS Trans Am Geophys Union* 70: 1.

Marty, B. and A. Jambon. 1987. C/^{3}He in volatile fluxes from the solid Earth: Impli-cations for carbon geodynamics. *Earth Planet Sci Lett* 83: 16–26.

Mast, M. A. 1989. A laboratory and field study of chemical weathering with special reference to acid deposition [Ph.D. dissertation]. Laramie, WY: University of Wyoming.

Maynard Smith, J. 1988. "Evolutionary Progress and Levels of Selection." In: M. H. Nitecki, ed., *Evolutionary Progress,* pp. 219–30. Chicago: University of Chicago Press.

Maynard Smith, J. and E. Szathmary. 1995. *The Major Transitions in Evolution.* Ox-ford, England: W. H. Freeman.

Maynard Smith, J. and E. Szathmary. 1996. On the likelihood of habitable worlds. *Nature* 384: 107.

Mayor, M. and D. Queloz. 1995. A Jupiter-mass companion to a solar-type star. *Nature* 378: 355–9.

Mayor, M., D. Queloz, S. Udry, and J-L. Halbwachs. 1997. "From Brown Dwarfs to Planets." In: C. B. Cosmovici, S. Bowyer, and D. Werthimer, eds., *Astronomical and Biochemical Origins and the Search for Life in the Universe,* pp. 313–30. Bologna, Italy: Editrice Compositori.

Mayr, E. 1961. Cause and effect in biology. *Science* 134: 1501–16.

Mayr, E. 1985. "The Probability of Extraterrestrial Intelligent Life." In: E. Regis, Jr., ed., *Extraterrestrials,* pp. 23–30. Cambridge, England: Cambridge University Press.

McCulloch, M. T. and V. C. Bennett. 1994. Progressive growth of the Earth's continental crust and depleted mantle: Geochemical constraints. *Geochim Cosmochim Acta* 58: 4717–38.

McElwain, J. C. and W. G. Chaloner. 1995. Stomatal density and index of fossil plants track atmospheric carbon dioxide in the Palaeozoic. *Ann Botany* 76: 389–95.

McGhee, G. R. 1996. *The Late Devonian Mass Extinction: The Frasnian/Famennian Crisis.* New York: Columbia University Press.

McIntyre, D. B. 1963. "James Hutton and Philosophy of Geology." In: C. C. Albritton Jr., ed., *The Fabric of Geology,* pp. 1–11. Stanford, CA: Freeman, Cooper, & Company.

McKay, C. P. and W. L. Davis. 1991. Duration of liquid water habitats on early Mars. *Icarus* 90: 214–21.

McKay, C. P., R. D. Lorenz, and J. I. Lunine. 1999. Analytic solutions for the antigreenhouse effect: Titan and the early Earth. *Icarus* 137: 56–61.

McKinnon, W. B. 1997. Extreme cratering. *Science* 276: 1346–8.

McMenamin, M. A. S. and D. L. S. McMenamin. 1990. *The Emergence of Animals. The Cambrian Breakthrough.* New York: Columbia University Press.

McMenamin, M. A. S. and D. L. S. McMenamin. 1993. Hypersea and the land ecosystem. *BioSystems* 31: 145–53.

McMenamin, M. A. S. and D. L. S. McMenamin. 1994. *Hypersea.* New York: Columbia University Press.

Merino, E. 1992, Self-organization in stylolites. *Am Scientist* 80: 466–73.

Michalopoulos, P. and R. C. Aller. 1995. Rapid clay mineral formation in Amazon delta sediments: Reverse weathering and oceanic elemental cycles. *Science* 270: 614–7.

Miller, A. R., R. S. Needham, and P. G. Stuart-Smith. 1992. "Mineralogy and Geochemistry of the Pre-1.65 Ga Paleosol Under Kombolgie Formation Sandstone of the Pine Creek Geosyncline, Northern Territory, Australia." In: M. Schidlowski, S. Golubic, M. M. Kimberly, D. M. McKirdy, and P. A. Trudinger, eds., *Early Organic Evolution: Implications for Mineral and Energy Resources,* pp. 76–105. Berlin: Springer-Verlag.

Molnar, P. and P. England. 1990. Late Cenozoic uplift of mountain ranges and global climate change: chicken or egg? *Nature* 346: 29–34.

Mora, C. I., S. G. Driese, and L. A. Colarusso. 1996. Middle to late Paleozoic atmospheric CO_2 levels from soil carbonate and organic matter. *Science* 271: 1105–7.

Morowitz, H. J. 1978. *Foundations of Bioenergetics.* New York: Academic Press.

Morowitz, H. J. 1989. "Biology of a Cosmological Science." In: J. B. Callicott and R. T. Ames, eds., *Nature in Asian Traditions of Thought,* pp. 37–49. Albany, NY: SUNY Press.

Morowitz, H. J. 1992. *Beginnings of Cellular Life.* New Haven, CT: Yale University Press.

Morris, S. C. 1993. The fossil record and the early evolution of the Metazoa. *Nature* 361: 219–25.

Moss, A. J., P. Green, and J. Hutka. 1981. Static breakage of granitic detritus by ice and water in comparison with breakage by flowing water. *Sedimentology* 28: 261–72.

Moulton, K. L. and R. A. Berner. 1998. Quantification of the effect of plants on weathering: Studies in Iceland. *Geology* 26: 895–8.

Muehlenbachs, K. and R. N. Clayton. 1976. Oxygen isotope composition of the oceanic crust and its bearing on seawater. *J Geophys Res* 81: 4365–9.

Myneni, S. C. B., T. K. Tokunaga, and G. E. Brown Jr. 1997. Abiotic selenium redox transformations in the presence of Fe(II,III) oxides. *Science* 278: 1106–9.

Nahon, D. B. 1991. Self-organization in chemical laterite weathering. *Geoderma* 51: 5–13.

Nash, T. H. III, ed. 1996. *Lichen Biology.* Cambridge, England: Cambridge University Press.

National Academy of Sciences. 1983. *Changing Climate: Report of the Carbon Dioxide Assessment Committee.* Washington, DC: National Academy Press.

Nisbet, E. G. 1987. *The Young Earth: An Introduction to Archaean Geology.* Boston, MA: Allen & Unwin.

Nisbet, E. G. 1995. "Archaean Ecology: A Review of Evidence for the Early Development of Bacterial Biomes, and Speculations on the Development of a Global-scale Biosphere." In: M. P. Coward and A. C. Ries, eds., *Early Precambrian Processes*, pp. 27–51. GSA Special Publication No. 95. Boulder, CO: Geological Society of America.

Nishio, Y., S. Sasaki, T. Gamo, H. Hiyagon, and Y. Sano. 1998. Carbon and helium isotope systematics of North Fiji Basin basalt glasses: Carbon geochemical cycle in the subduction zone. *Earth Planet Sci Lett* 154: 127–38.

Noll, K. S., T. L. Roush, D. P. Cruikshank, R. E. Johnson, and Y. J. Pendleton. 1997. Detection of ozone on Saturn's satellites Rhea and Dione. *Nature* 388: 45–7.

Nugent, M. A., S. L. Brantley, C. G. Pantano, and P. A. Maurice. 1998. The influence of natural mineral coatings on feldspar weathering. *Nature* 395: 588–90.

Oberbeck, V. R. 1993. Impacts and global change. *Geotimes* (September): 16–8.

Oberbeck, V. R. and R. L. Mancinelli. 1994. Asteroid impacts, microbes, and the cooling of the atmosphere. *BioScience* 44: 174–7.

Oberbeck, V. R. and J. R. Marshall. 1992. Impacts, flood basalts, and continental breakup [Abstract]. *Lunar Planet Sci Proc* (March 23): 113.

Oberbeck, V. R., J. R. Marshall, and H. Aggarwal. 1993. Impacts, tillites, the breakup of Gondwanaland. *J Geol* 101: 1–19.

Ohmoto, H. 1996. Evidence in pre-2.2 Ga paleosols for the early evolution of atmospheric oxygen and terrestrial biota. *Geology* 24: 1135–8.

Ohmoto, H. 1997a. When did the Earth's atmosphere become oxic? *Geochem News* 93: 12, 13, 26, 27.

Ohmoto, H. 1997b. Evidence in pre-2.2 Ga paleosols for the early evolution of atmospheric oxygen and terrestrial biota: Reply. *Geology* 25: 858–9.

Ohmoto, H. and R. P. Felder. 1987. Bacterial activity in the warmer, sulphate-bearing, Archean oceans. *Nature* 328: 244–6.

Olendzanski, L. and J. P. Gogarten. (1998). "Deciphering the Molecular Record for the Early Evolution of Life: Gene Duplication and Horizontal Gene Transfer." In: J. Wiegel and M. W. W. Adams, eds., *Thermophiles: The Keys to Molecular Evolution and the Origin of Life?* pp. 165–76. London: Taylor and Francis.

Ortoleva, P. 1994. *Geochemical Self-organization.* New York: Oxford University Press.

Ortoleva, P., W. Merino, C. Moore, and J. Chadam. 1987. Geochemical self-organization I: Reaction-transport feedbacks and modeling approach. *Am J Sci* 287: 979–1007.

Owen, T. and A. Bar-Nun. 1995. Comets, impacts, and atmospheres. *Icarus* 116: 215–26.

Pace, N. R. 1997. A molecular view of microbial diversity and the biosphere. *Science* 276: 734–40.

Peixoto, J. P. and A. H. Oort. 1992. *Physics of Climate*. New York: American Institute of Physics.

Peixoto, J. P., A. H. Oort, M. De Almeida, and A. Tome. 1991. Entropy budget of the atmosphere. *J Geophys Res* 96: 10,981–8.

Pennesi, E. 1998. Genome data shake the tree of life. *Science* 280: 672–4.

Perry, E. C. 1990. Comment on "The implication of the oxygen isotope records in coexisting cherts and phosphates." *Geochim Cosmochim Acta* 54: 1175–9.

Perry, E. C. and F. C. Tan. 1972. Significance of oxygen and carbon isotope variations in early Precambrian cherts and carbonate rocks of southern Africa. *Geol Soc Am Bull* 83: 647–64.

Perry, E. C. Sr. 1967. The oxygen isotopic chemistry of ancient cherts. *Earth Planet Sci Lett* 3: 62–6.

Petrovich, R. 1981. Kinetics of dissolution of mechanically comminuted rock-forming oxides and silicates I. Deformation and dissolution of quartz under laboratory conditions. *Geochim Cosmochim Acta* 45: 1665–74.

Petrovich, R. 1981. Kinetics of dissolution of mechanically comminuted rock-forming oxides and silicates II. Deformation and dissolution of oxides in the laboratory and at the Earth's surface. *Geochim Cosmochim Acta* 45: 1675–86.

Pierson, B. K. 1994. "The Emergence, Diversification, and Role of Photosynthetic Eubacteria." In: S. Bengtson, ed., *Early Life on Earth*, Nobel Symposium No. 84. pp. 161–80. New York: Columbia University Press.

Pineau, F. and M. Javoy. 1994. Strong degassing at ridge crests: The behavior of dissolved carbon and water in basalt glasses at 14°N, Mid-Atlantic Ridge. *Earth Planet Sci Lett* 123: 179–98.

Plummer, L. N. and E. Busenberg. 1982. The solubilities of calcite, aragonite and vaterite in CO_2-H_2O solutions between 0 and 90°C, and an evaluation of the aqueous model for the system $CaCO_3$-CO_2-H_2O. *Geochim Cosmochim Acta* 46: 1011–40.

Pollack, J. B., J. F. Kasting, S. M. Richardson, and K. Poliakoff. 1987. The case for a wet, warm climate on early Mars. *Icarus* 71: 203–24.

Polynov, B. B. 1945. The first stages of soil formation on massive-crystalline rocks. *Pochvovedenie* No. 7, pp. 327–39 (translation: Washington, DC: Office of Technical Services, U.S. Department of Commerce, IPST Cat. No. 1350).

Polynov, B. B. 1948. Leading ideas in the present-day theory of soil formation and soil development. *Pochvovedenie* No. 1, pp. 3–13 (translation: Washington, DC:

Office of Technical Services, U.S. Department of Commerce, IPST Cat. No. 1351).

Polynov, B. B. 1953. The geological role of organisms. *Voprosy Geogr* Collection 33: 45–64 (translation: Washington, DC: Office of Technical Services, U.S. Department of Commerce, IPST Cat. No. 1361).

Raff, R. A. 1996. *The Shape of Life*. Chicago: University of Chicago Press.

Rampino, M. R. 1992. Ancient "glacial" deposits are ejecta of large impacts: The ice age paradox explained [Abstract A32C-1]. *EOS Trans Am Geophys Union* 73 (Suppl): 99.

Rampino, M. R. 1994. Tillites, diamictites, and ballistic ejecta of large impacts. *J Geol* 102: 439–56.

Rampino, M. R., K. Ernstson, and F. Anguita. 1997. Striated and polished clasts in impact-ejecta and the "tillite problem." Geological Society of America Abstracts with Programs 29, No.7, p. A-81.

Rampino, M. R., B. M. Haggerty, and T. C. Pagano. 1997. A unified theory of impact crises and mass extinctions: Quantitative tests. *Ann N Y Acad Sci* 822: 403–31.

Rampino, M. R., D. W. Schwartzman, K. Caldeira, and P. D. Schwartzman. 1996. Impacts and Precambrian climate: The Huronian enigma [Abstract]. 5th International Conference on Bioastronomy, IAU Colloquium No. 161, Capri, July 1–5, 1996.

Raymo, M. E. 1991. Geochemical evidence supporting T. C. Chamberlin's theory of glaciation. *Geology* 19: 344–7.

Raymo, M. E. 1994. The Himalayas, organic burial, and climate in the Miocene. *Paleoceanography* 9: 399–404.

Raymo, M. E. and W. F. Ruddiman. 1992. Tectonic forcing of late Cenozoic climate. *Nature* 359: 117–22.

Raymo, M. E., W. F. Ruddiman, and P. N. Froelich. 1988. Influence of late Cenozoic mountain building on ocean geochemical cycles. *Geology* 16: 649–53.

Retallack, G. J. 1990. *Soils of the Past: An Introduction to Paleopedology*. Boston: Unwin Hyman.

Retallack, G. J. 1997. Early forest soils and their role in Devonian global change. *Science* 276: 583–5.

Reusch, D. N. and K. A. Maasch. 1995. Weathering of simatic crust, the carbon cycle, and climate [Abstract U42B-9]. *EOS Trans Am Geophys Union* 76: S53.

Richter, D. D. and D. Markewitz. 1995. How deep is soil? *BioScience* 45: 600–9.

Robbins, E. I., K. G. Porter, and K. A. Haberyan. 1985. Pellet microfossils: Possible evidence for metazoan life in early Proterozoic time. *Proc Natl Acad Sci U S A* 82: 5809–13.

Robert, M. and J. Berthelin. 1986. "Role of Biological and Biochemical Factors in Soil Mineral Weathering." In: P. M. Huang and M. Schnitzer, eds., *Interactions of Soil Minerals with Natural Organics and Microbes,* pp. 453–9. Special Publication Vol. 17. Madison, WI: Soil Science Society of America.

Robinson, J. M. 1991. Land plants and weathering. *Science* 252: 860.

Rodhe, H. 1992. "Modeling Biogeochemical Cycles." In: S. S. Butcher, R. J. Charlson, G. H. Orians, and G. V. Wolfe, eds., *Global Biogeochemical Cycles,* pp. 55–72. London: Academic Press.

Rothschild, L. J. and R. L. Mancinelli. 1990. Model of carbon fixation in microbial mats from 3,500 Myr ago to the present. *Nature* 345: 710–2.

Runnegar, B. 1991. Precambrian oxygen levels estimated from the biochemistry and physiology of early eukaryotes. *Palaeogeogr Palaeoclimatol Palaeoecol* (Global Planetary Change Section) 97: 97–111.

Russell, D. A. 1981. "Speculations on the Evolution of Intelligence in Multicellular Organisms." In: J. Billingham, ed., *Life in the Universe,* pp. 259–75. Cambridge, MA: MIT Press.

Russell, M. J., R. M. Daniel, A. J. Hall, and J. A. Sherringham. 1994. A hydrothermally precipitated catalytic iron sulfide membrane as a first step to life. *J Mol Evolution* 39: 231–43.

Rye, R., P. H. Kuo, and H. D. Holland. 1995. Atmospheric carbon dioxide concentrations before 2.2 billion years ago. *Nature* 378: 603–5.

Sackmann, I-J., A. I. Boothroyd, and W. A. Fowler. 1990. Our Sun. I. The standard model: Successes and failures. *Astrophys J* 360: 727–36.

Sagan, C. 1977. Reducing greenhouses and the temperature history of Earth and Mars. *Nature* 269: 224–6.

Sagan, C. and C. Chyba. 1997. The early faint sun paradox: Organic shielding of ultraviolet-labile greenhouse gases. *Science* 276: 1217–21.

Sagan, C. and G. Mullen. 1972. Earth and Mars: Evolution of atmospheres and surface temperatures. *Science* 177: 52–6.

Salop, L. J. 1983. *Geological Evolution of the Earth during the Precambrian.* New York: Springer-Verlag.

Sano, Y. and S. N. Williams. 1996. Fluxes of mantle and subducted carbon along convergent plate boundaries. *Geophys Res Lett* 23: 2749–52.

Santhyendranath, S., A. D. Gouveia, S. R. Shetye, P. Ravindran, and R. Platt. 1991. Biological control of surface temperature in the Arabian Sea. *Nature* 349: 54–6.

Saunders, P. T. 1994. Evolution without natural selection: Further implications of the Daisyworld parable. *J Theor Biol* 166: 365–73.

Schidlowski, M. 1988. A 3800-million-year isotopic record of life from carbon in sedimentary rocks. *Nature* 333: 313–8.

Schidlowski, M. and P. Aharon 1992. "Carbon Cycle and Carbon Isotope Record: Geochemical Impact of Life over 3.8 Ga of Earth History." In: M. Schidlowski, S. Golubic, M. M. Kimberly, D. M. McKirdy, and P. A. Trudinger, eds., *Early Organic Evolution: Implications for Mineral and Energy Resources*, pp. 147–75. Berlin: Springer-Verlag.

Schidlowski, M., J. M Hayes, and I. R. Kaplan. 1983. "Isotopic Inferences of Ancient Biochemistries: Carbon, Sulfur, Hydrogen, and Nitrogen. In: J. W. Schopf, ed., *Earth's Earliest Biosphere*, pp. 149–86. Princeton, NJ: Princeton University Press.

Schneider, S. H. and P. Boston, eds. 1991. *Scientists on Gaia*. Cambridge, MA: MIT Press.

Schneider, S. H. and R. Londer. 1984. *The Coevolution of Climate and Life*. San Francisco: Sierra Club Books.

Schopf, J. W. 1992. "Paleobiology of the Archean." In: J. W. Schopf and C. Klein, eds., *The Proterozoic Biosphere*, pp. 25–39. Cambridge, England: Cambridge University Press.

Schopf, J. W. 1993. Microfossils of the early Archean apex chert: New evidence of the antiquity of life. *Science* 260: 640–6.

Schultz, P. H. and D. E. Gault. 1990. "Prolonged Global Catastrophes from Oblique Impacts." Special Paper 247. In: V. L. Sharpton and P. D. Ward, eds., *Global Catastrophes in Earth History: An Interdisciplinary Conference on Impacts, Volcanism, and Mass Mortality*, pp. 239–61. Boulder, CO: Geological Society of America.

Schumm, S. A. 1968. Speculations concerning paleohydrologic controls on terrestrial sedimentation. *Geol Soc Am Bull* 79: 1573–88.

Schwartzman, D. 1993. Comment on "Weathering, plants, and the long-term carbon cycle" by Robert A. Berner. *Geochim Cosmochim Acta* 57: 2145–6.

Schwartzman, D. 1994a. Biotic enhancement of weathering redux. Goldschmidt Conference Abstracts, Edinburgh. *Mineralogic Mag* 58A: 815–6.

Schwartzman, D. 1994b. "Temperature and the Evolution of the Biosphere." In: G. Seth Shostak, ed., *Progress in the Search for Extraterrestrial Life. 1993 Bioastronomy Symposium, Santa Cruz, California*, pp. 153–61. San Francisco: Astronomical Society of the Pacific.

Schwartzman, D. and K. Caldeira. 1995. Rethinking the sedimentary carbon isotopic record. The lst SEPM Congress on Sedimentary Geology, August 13–16, St. Pete Beach, FL. Congress Program and Abstracts, Vol. 1, p. 111.

Schwartzman, D. and M. McMenamin. 1993. A much warmer Earth surface for most of geologic time: Implications to biotic weathering. *Chem Geol* 107: 221–3.

Schwartzman, D. and L. J. Rickard. 1988. Being optimistic about the search for extraterrestrial intelligence. *Am Sci* 76: 364–9.

Schwartzman, D. and S. Shore. 1996. Biotically mediated surface cooling and habitability for complex life. In: L. R. Doyle, ed., *Circumstellar Habitable Zones*, pp. 421–43. Menlo Park, CA: Travis House Publishers.

Schwartzman, D. and T. Volk. 1991a. Biotic enhancement of weathering and surface temperatures on Earth since the origin of life. *Palaeogeogr Palaeoclimatol Palaeoecol* (Global Planetary Change Section) 90: 357–71.

Schwartzman, D. and T. Volk. 1991b. "Geophysiology and Habitable Zones Around Sun-like Stars." In: J. Heidmann and M. J. Klein, eds., *Bioastronomy, Proceedings, Val Cenis, France, 1990*, pp. 155–62. Lecture Notes in Physics 390. New York: Springer-Verlag.

Schwartzman, D. and T. Volk. 1992. "Biotic Enhancement of Earth Habitability." In: W. A. Nierenberg, ed., *Encyclopedia of Earth System Science*, Vol. 1, pp. 387–94. San Diego, CA: Academic Press.

Schwartzman, D. W. 1975. Althusser, dialectical materialism and the philosophy of science. *Sci Soc* 39: 318–30.

Schwartzman, D. W. 1991. Lichens as monitors of heavy metals and agents of weathering. Final Report, National Geographic Society Grant 4021–89, June 14, 1991.

Schwartzman, D. W. 1998. "Life Was Thermophilic for the First Two-thirds of Earth History." In: J. Wiegel and M. W. W. Adams, eds., *Thermophiles: The Keys to Molecular Evolution and the Origin of Life?* pp. 33–43. London: Taylor and Francis.

Schwartzman, D. W., R. Aghamiri, and S. W. Bailey. 1997. Lichen weathering rates from miniwatershed studies. Geological Society of America, Abstracts with Programs 29, No.7, p. A-361.

Schwartzman, D., M. McMenamin, and T. Volk. 1993. Did surface temperatures constrain microbial evolution? *BioScience* 43: 390–3.

Schwartzman, D., S. Shore, T. Volk, and M. McMenamin. 1994. Self-organization of the Earth's biosphere—Geochemical or geophysiological? *Origins Life Evolution Biosphere* 24: 435–50.

Schwartzman, D. W. and T. Volk. 1989. Biotic enhancement of weathering and the habitability of Earth. *Nature* 340: 457–60.

Schwartzman, D. W. and T. Volk. 1990. From abiotic to biotic Earth: The carbon cycle's climatic consequences. V. M. Goldschmidt Conference, Program and Abstracts, May 2–4, Baltimore, MD, The Geochemical Society, p. 80.

Seckbach, J. 1994. The first eukaryotic cells—Acid hot-spring algae. *J Biol Phys* 20: 335–45.

Seckbach, J. 1997. "Search for Life in the Universe with Terrestrial Microbes which Thrive under Extreme Conditions." In: C. B. Cosmovici, S. Bowyer, and D. Werthimer, eds., *Astronomical and Biochemical Origins and the Search for Life in the Universe*, pp. 511–23. Bologna, Italy: Editrice Compositori.

Seilacher, A. 1997. Precambrian life styles related to biomats. Geological Society of America Abstracts with Programs 29, No.7, p. A-193.

Seilacher, A., P. K. Bose, and F. Pfluger. 1998. Triploblastic animals more than 1 billion years ago: Trace fossil evidence from India. *Science* 282: 80–3.

Sellers, W. D. 1974. *Physical Climatology*. Chicago: University of Chicago Press.

Serafin, R. 1988. Noosphere, Gaia, and the science of the biosphere. *Environ Ethics* 10: 121–37.

Sharp, M., M. Tranter, G. H. Brown, and M. Skidmore. 1995. Rates of chemical denudation and CO_2 drawdown in a glacier-covered alpine catchment. *Geology* 23: 61–4.

Shaw, G. E. 1987. Aerosols as climate regulators: A climate-biosphere linkage? *Atmospher Environ* 21: 985–6.

Sherrif, B. L. and D. Brown. 1995. Microbial geochemistry of granite. Annual Meeting, Geological Society of America, Abstracts with Programs 27, No. 6, p. A-185.

Shklovskii, I. S. and C. Sagan. 1966. *Intelligent Life in the Universe*. San Francisco: Holden-Day.

Simpson, G. G. 1964. *This View of Life*. New York: Harcourt, Brace and World.

Skidmore, M. L., J. Foght, and M. J. Sharp. 1997. Micorbially mediated weathering reactions in glacial environments. Geological Society of America Abstracts with Programs 29, No. 7, p. A-362.

Skinner, B. J. and S. C. Porter. 1995. *The Blue Planet: An Introduction to Earth System Science*. New York: Wiley and Sons.

Smolin, L. 1997. *The Life of the Cosmos*. New York: Oxford University Press.

Sogin, M. L. 1997. History assignment: When was the mitochondrion founded? *Cur Opin Genet Devel* 7: 792–9.

Sogin, M. L., J. H. Gunderson, H. J. Elwood, D. A. Alonso, and D. A. Peattie. 1989. Phylogenetic meaning of the kingdom concept: An unusual ribosomal RNA from Guardia lamlia. *Science* 243: 75–7.

Sommaruga-Wograth, S., K. A. Koinig, R. Schmidt, R. Sommaruga, R. Tessadri, and R. Psenner. 1997. Temperature effects on the acidity of remote alpine lakes. *Nature* 387: 64–7.

Southard, J. B. and L. A. Bouguchwal. 1990a. Bed configurations in steady unidirectional water flows. Part 2. Synthesis of flume data. *J Sed Petrol* 60: 658–79.

Southard, J. B. and L. A. Bouguchwal. 1990b. Bed configurations in steady unidirectional water flows. Part 3. Effects of temperature and gravity. *J Sed Petrol* 60: 680–6.

Spencer, R. J. and L. A. Hardie. 1990. "Control of Seawater Composition by Mixing of River Waters and Mid-ocean Ridge Hydrothermal Brines." In: R. J. Spencer and I-Ming Chou, eds., *Fluid-Mineral Interactions: A Tribute to H. P. Eugster.* Special Publication No. 2. Columbus, OH: Geochemical Society.

Spivack, A. J. and H. Staudigel. 1994. Low-temperature alteration of the upper oceanic crust and the alkalinity budget of seawater. *Chem Geol* 115: 239–47.

Spooner, E. T. C. 1992. Similarities between environmental requirements for the deepest known branches of the universal phylogenetic tree and early Archean (~3.0–3.5 Ga) whole ocean conditions [Abstract]. Program of the Annual Meeting of the Geological Society of America 24, No. 7, p. A137.

Stahlecher, E. 1906. Untersuchungen über Thallusbau in ihren Beziehungen zum Substrat bei siliciseden Krustenflechten. *Fünfstüks Beitr Wiss Bot* 5: 405–51.

Stallard, R. F. 1992. "Tectonic Processes, Continental Freeboard, and the Rate-controlling Step for Continental Denudation." In: S. S. Butcher, R. J. Charlson, G. H. Orians, and G. V. Wolfe, eds., *Global Biogeochemical Cycles,* pp. 93–121. London: Academic Press.

Stallard, R. F. 1995a. "Relating Chemical and Physical Erosion." In: A. F. White and S. L. Brantley, eds., *Chemical Weathering Rates of Silicate Minerals. Reviews in Mineralogy,* Vol. 31, pp. 543–64. Washington, DC: Mineralogical Society of America.

Stallard, R. F. 1995b. Tectonic, environmental, and human aspects of weathering and erosion: A global review using a steady-state perspective. *Ann Rev Earth Planet Sci* 23: 11–39.

Staudigel, H., R. A. Chastain, A. Yayanos, and W. Bourcier. 1995. Biologically mediated dissolution of glass. *Chem Geol* 126: 147–54.

Staudigel, H., S. R. Hart, H-U. Schmincke, and B. M. Smith. 1989. Cretaceous ocean crust at DSDP Sites 417 and 418: Carbon uptake from weathering versus loss by magmatic outgassing. *Geochim Cosmochim Acta* 53: 3091–4.

Staudigel, H., S. R. Hart, H-U. Schmincke, and B. M. Smith. 1990. Reply to

"Global CO_2 degassing and the carbon cycle": A comment by R. A. Berner. *Geochim Cosmochim Acta* 54: 2891.

Steefel, C. I. and P. Van Cappellen. 1990. A new kinetic approach to modeling water- rock interaction: The role of nucleation, precursors and Ostwald ripening. *Geochim Cosmochim Acta* 54: 2657–77.

Stevens, T. O. and J. P. McKinley. 1995. Lithoautotrophic microbial ecosystems in deep basalt aquifers. *Science* 270: 450–4.

Strauss, H., D. J. Des Marais, J. M. Hayes, and R. E. Summons. 1992a. "The Carbon-isotopic Record." In: J. W. Schopf and C. Klein, eds., *The Proterozoic Biosphere*, pp. 117–27. Cambridge, England: Cambridge University Press.

Strauss, H., D. J. Des Marais, J. M. Hayes, and R. E. Summons. 1992b. "Proterozoic Organic Carbon—Its Preservation and Isotopic Record." In: M. Schidlowski, S. Golubic, M. M. Kimberly, D. M. McKirdy, and P. A. Trudinger, eds., *Early Organic Evolution: Implications for Mineral and Energy Resources*, pp. 203–11. Berlin: Springer-Verlag.

Strother, P. K., S. Al-Hajri, and A. Traverse. 1996. New evidence for land plants from the lower Middle Ordovician of Saudi Arabia. *Geology* 24: 55–8.

Summerfield, M. A. 1991. *Global Geomorphology.* New York: Wiley and Sons.

Summers, D. P. and S. Chang. 1993. Prebiotic ammonia from reduction of nitrite by iron (II) on the early Earth. *Nature* 365: 630–3.

Sumner, D. Y. and J. P. Grotzinger. 1996. Were kinetics of Archean calcium carbonate precipitation related to oxygen concentration? *Geology* 24: 119–22.

Sundquist, E. T. 1991. Steady- and non-steady-state carbonate-silicate controls on atmospheric CO_2. *Quaternary Sci Rev* 10: 283–96.

Sverdrup, H. 1990. *The Kinetics of Base Cation Release Due to Chemical Weathering.* Lund: Lund University Press.

Sverdrup, H. and P. Warfvinge. 1988. Weathering of primary silicate minerals in the natural soil environment in relation to a chemical weathering model. *Water Air Soil Pollution* 38: 387–408.

Sverdrup, H. and P. Warfvinge. 1995. "Estimating Field Weathering Rates Using Laboratory Kinetics." In: A. F. White and S. L. Brantley, eds., *Chemical Weathering Rates of Silicate Minerals. Reviews in Mineralogy,* Vol. 31, pp. 485–541. Washington, DC: Mineralogical Society of America.

Sverdrup, H. U., M. W. Johnson, and R. H. Fleming. 1942. *The Oceans: Their Physics, and General Biology.* New York: Prentice-Hall.

Swenson, R. 1990. "The Earth as an Incommensurate Field at the Geo-cosmic Interface: Fundamentals to a Theory of Emergent Evolution." In: G.J.M. Tomassen,

W. de Graaff, A. A. Knoop and R. Hengeveld, eds., *Geo-cosmic Relations; the Earth and its Macro-Environment,* pp. 299–306. Wageningen, The Netherlands: Pudoc.

Swenson, R. and M. T. Turvey. 1991. Thermodynamic reasons for perception-action cycles. *Ecol Psychol* 3: 317–48.

Syers, J. K. and I. K. Iskandar. 1973. "Pedogenic Significance of Lichens." In: V. Ahmadjian and M. E. Hale Jr., eds., *The Lichens,* pp. 225–48. New York: Academic Press.

Sylvester, P. J. 1998. Formation of the continents—dribble or big bang? *Geochem News* 94: 12, 13, 23–25.

Sylvester, P. J., I. H. Campbell, and D. A. Bowyer. 1997. Niobium/uranium evidence for early formation of the continental crust. *Science* 275: 521–3.

Sylvester, P. J., J. C. Dann, and M. G. Green. 1998. Nb/U of 3.5 Ga basalts, depleted mantle, and continental growth. *Geological Society of America, Abstracts with Programs.* Annual Meeting. Toronto. A–207.

Tajika, E. and T. Matsui. 1990. "The Evolution of the Terrestrial Environment." In: H. E. Newsom and J. H. Jones, eds., *Origin of the Earth,* pp. 347–70. New York: Oxford University Press.

Tajika, E. and T. Matsui. 1992. Evolution of the terrestrial proto-CO_2 atmosphere coupled with thermal history of the earth. *Earth Planet Sci Lett* 113: 251–66.

Tajika, E. and T. Matsui. 1993. Degassing history and carbon cycle of the Earth: From an impact-induced steam atmosphere to the present atmosphere. *Lithos* 30: 267–80.

Taylor, A. B. and M. A. Velbel. 1991. Geochemical mass balance and weathering rates in forested watersheds of the southern Blue Ridge. II. Effects of botanical uptake terms. *Geoderma* 51: 29–50.

Taylor, S. R. and S. M. McLennan. 1995. The geochemical evolution of the continental crust. *Rev Geophys* 33: 241–65.

Taylor, T. N., H. Hass, W. Remy, and H. Kerp. 1995. The oldest fossil lichen. *Nature* 378: 244.

Thomas, M. F. 1994. *Geomorphology in the Tropics.* Chichester, England: Wiley and Sons.

Thomson, J. W. 1984. *American Arctic Lichens.* Vol. 1. The Macrolichens. New York: Columbia University Press.

Thornes, J. B., ed. 1990. *Vegetation and Erosion: Processes and Environments.* Chichester, England: Wiley & Sons.

Thornton, I. W. B. 1996. *Krakatau: the Destruction and Reassembly of an Island Eco-System.* Cambridge, MA: Harvard University Press.

Thorseth, I. H., H. Furnes, and M. Heldal. 1992. The importance of microbiological activity in the alteration of natural basaltic glass. *Geochim Cosmochim Acta* 56: 845–50.

Thorseth, I. H., H. Furnes, and O. Tumyr. 1995a. Textural and chemical effects of bacterial activity on basaltic glass: An experimental approach. *Chem Geol* 119: 139–60.

Thorseth, I. H., T. Torsvik, H. Furnes, and K. Muehlenbachs. 1995b. Microbes play an important role in the alteration of oceanic crust. *Chem Geol* 126: 137–46.

Tipler, F. J. 1981. Additional remarks on extraterrestrial intelligence. *Q J R Astron Soc* 22: 279–92.

Towe, K. M. 1970. Oxygen-collagen priority and the early metazoan fossil record. *Proc Natl Acad Sci U S A* 65: 781–8.

Towe, K. M. 1981. Biochemical keys to the emergence of complex life. In: J. Billingham, ed., *Life in the Universe*, pp. 297–307. Cambridge, MA: MIT Press.

Towe, K. M. 1994. "Earth's Early Atmosphere: Constraints and Opportunities for Early Evolution." In: S. Bengtson, ed., *Early Life on Earth*, Nobel Symposium No. 84. pp. 36–47. New York: Columbia University Press.

Treub, M. 1888. Notice sur la nouvelle flore de Krakatau. *Ann Jardin Botanique Buitenzorg* 7: 213–23.

Trumbore, S. E., Chadwick, O. A., and R. Amundson. 1996. Rapid exchange between soil carbon and atmospheric carbon dioxide driven by temperature change. *Science* 272: 393–6.

Twist, D. and E. S. Cheney. 1986. Evidence for the transition to an oxygen-rich atmosphere in the Rooiberg Group, South Africa—A note. *Precambrian Res* 33: 255–64.

Ulanowicz, R. E. and Hannon, B. M. 1987, Life and the production of entropy. *Proc R Soc Lond* B232: 181–92.

Urey, H. C. 1952. *The Planets: Their Origin and Development.* New Haven, CT: Yale University Press.

Veizer, J. 1988. "The Evolving Exogenic Cycle." In: C. B. Gregor, R. M. Garrels, F. T. Mackenzie, J. B. Maynard, eds., *Chemical Cycles in the Evolution of the Earth*, pp. 175–220. New York: Wiley and Sons.

Veizer, J. 1994. "The Archean-Proterozoic Transition and Its Environmental Implications." In: S. Bengtson, ed., *Early Life on Earth*, Nobel Symposium No. 84. pp. 208–19. New York: Columbia University Press.

Veizer, J., R. N. Clayton, and R. W. Hinton. 1992. Geochemistry of Precambrian carbonates: IV. Early Paleoproterozoic (2.25 ± 0.25 Ga) seawater. *Geochim Cosmochim Acta* 56: 875–85.

Veizer, J., J. Hoefs, D. R. Lowe, and P. C. Thurston. 1989b. Geochemistry of Precambrian carbonates: II. Archean greenstone belts and Archean sea water. *Geochim Cosmochim Acta* 53: 859–71.

Veizer, J., J. Hoefs, R. H. Ridler, L. S. Jensen, and D. R. Lowe. 1989a. Geochemistry of Precambrian carbonates: I. Archean hydrothermal systems. *Geochim Cosmochim Acta* 53: 845–57.

Velbel, M. A. 1990. Influence of temperature and mineral surface characteristics on feldspar weathering rates in natural and artificial systems: A first approximation. *Water Resources Res* 26: 3049–53.

Velbel, M. A. 1993. Temperature dependence of silicate weathering in nature: How strong a negative feedback on long-term accumulation of atmospheric CO_2 and global greenhouse warming? *Geology* 21: 1059–62.

Velbel, M. A. 1995. "Interactions of Ecosystem Processes and Weathering Processes." In: S. Trudgill, ed., *Solute Modelling in Catchment Systems*, pp. 193–209. New York: Wiley and Sons.

Vernadsky, V. I. 1998. *The Biosphere*. New York: Copernicus, Springer-Verlag.

Vernadsky, W. I. 1944. Problems of biogeochemistry, II. *Trans Conn Acad Arts Sci* 35: 483–517.

Vernadsky, W. I. 1945. The biosphere and the noosphere. *Am Sci* 33: 1–12.

Vervoort, J. D., P. J. Patchett, G. E. Gehrels, and A. P. Nutman. 1996. Constraints on early Earth differentiation from hafnium and neodymium isotopes. *Nature* 379: 624–7.

Volk, T. 1987. Feedbacks between weathering and atmospheric CO_2 over the last 100 million years. *Am J Sci* 287: 763–79.

Volk, T. 1989a. Sensitivity of climate and atmospheric CO_2 to deep-ocean and shallow-ocean carbonate burial. *Nature* 337: 637–40.

Volk, T. 1989b. Rise of angiosperms as a factor in long-term cooling. *Geology* 17: 107–10.

Volk, T. 1993. Cooling in the late Cenozoic. *Nature* 361: 123.

Volk, T. 1994. The soil's breath. *Natural History* 103: 48–54.

Volk, T. 1998. *Gaia's Body: Toward a Physiology of Earth*. New York: Copernicus, Springer-Verlag.

Wachtershauser, G. 1988. Before enzymes and templates: Theory of surface metabolism. *Microbiol Rev* 52: 452–84.

Wachtershauser, G. 1998. The case for a hyperthermophilic, chemolithoautotrophic origin of life in an iron-sulfur world. Chapter 4. In: J. Wiegel and M. Adams, eds., *Thermophiles: The Keys to Molecular Evolution and the Origin of Life?* pp. 47–57. London: Taylor and Francis.

Waldrop, M. M. 1990. Spontaneous order, evolution and life. *Science* 247: 1543–5.

Walker, J. C. G. 1977. *Evolution of the Atmosphere*. New York: MacMillan.

Walker, J. C. G. 1982. Climatic factors on the Archean Earth. *Palaeogeogr Palaeoclimatol Palaeoecol* 40: 1–11.

Walker, J. C. G. 1983. Possible limits on the composition of the Archaean ocean. *Nature* 302: 518–20.

Walker, J. C. G. 1985. Carbon dioxide on the early Earth. *Origins Life* 16: 117–27.

Walker, J. C. G. 1990. Precambrian evolution of the climate system. *Palaeogeogr Palaeoclimatol Palaeoecol* (Global Planetary Change Section) 82: 261–89.

Walker, J. C. G., P. B. Hays, and J. F. Kasting. 1981. A negative feedback mechanism for the long-term stabilization of Earth's surface temperature. *J Geophys Res* 86: 9776–82.

Walker, J. C. G. and K. J. Zahnle. 1986. Lunar nodal tides and distance to the Moon during the Precambrian. *Nature* 370: 600–2.

Walter, M. R. and D. J. Des Marais. 1993. Preservation of biological information in thermal spring deposits: Developing a strategy for the search for fossil life on Mars. *Icarus* 101: 129–43.

Watson, A. J. and J. E. Lovelock. 1983. Biological homeostasis of the global environment: The parable of Daisyworld. *Tellus* 35B: 284–9.

Westbroek, P. 1991. *Life as a Geological Force*. New York: W. W. Norton.

Wetherill, G. W. 1996. The formation and habitability of extra-solar planets. *Icarus* 119: 219–38.

White, A. F. 1995. "Chemical Weathering Rates of Silicate Minerals in Soils." In: A. F. White and S. L. Brantley, eds., *Chemical Weathering Rates of Silicate Minerals. Reviews in Mineralogy*, Vol. 31, pp. 407–61. Washington, DC: Mineralogical Society of America.

White, A. F. and A. E. Blum. 1995. Effects of climate on chemical weathering in watersheds. *Geochim Cosmochim Acta* 59: 1729–47.

White, A. F. and S. L. Brantley, eds. 1995. *Chemical Weathering Rates of Silicate Minerals. Reviews in Mineralogy*, Vol. 31. Washington, DC: Mineralogical Society of America.

White, A. F., T. D. Bullen, D. V. Vivit, and M. S. Schulz. 1997. "Experimental Weathering of Granitoids: Importance of Relative Mineral Reaction Rates." LPI Contribution No. 921. In: *Seventh Annual V.M. Goldschmidt Conference*, pp. 217–8. Houston: Lunar and Planetary Institute.

White, A. F. and M. L. Peterson. 1990. "Role of Reactive-surface-area Characterization in Geochemical Kinetic Models." In: D.C. Melchior and R. L. Bassett, eds., *Chemical Modeling of Aqueous Systems II*, pp. 461–75. ACS Symposium Series 416. Washington, DC: American Chemical Society.

Whitmire, D. P., L. R. Doyle, R. T. Reynolds, and J. J. Matese. 1995. A slightly more massive Sun as an explanation for warm temperatures on early Mars. *J Geophys Res* 100: 5457–64.

Williams, D. M., J. F. Kasting, and R. A. Wade. 1997. Habitable moons around extrasolar giant planets. *Nature* 385: 234–6.

Williams, G. C. 1992. *Gaia*, nature worship and biocentric fallacies. *Q Rev Biol* 67: 479–86.

Williams, G. R. 1996. *The Molecular Biology of Gaia*. New York: Columbia University Press.

Willson, L. A., G. H. Bowen, and C. Struck-Marcell. 1987. Mass loss on the main sequence. *Comments Astrophysics* 12: 17–34.

Wilson, M. J. 1995. Interactions between lichens and rocks: A review. *Cryptogamic Botany* 5: 299–305.

Wilson, M. J. and D. Jones. 1983. "Lichen Weathering of Minerals and Implications for Pedogenesis." In: R.C.L. Wilson, ed., *Residual Deposits: Surface Related Weathering Processes and Materials,* pp. 5–12. Oxford, England: Blackwell Scientific.

Winter, B. L. and L. P. Knauth. 1992. Stable isotope geochemistry of early Proterozoic carbonate concretions in the Animikie Group of the Lake Superior region: Evidence for anaerobic bacterial processes. *Precambrian Res* 54: 131–51.

Woese, C. R. 1987. Bacterial evolution. *Microbiol Rev* 51: 221–71.

Worsley, T. R. and D. L. Kidder. 1991. First-order coupling of paleogeography and CO_2, with global surface temperature and its latitudinal contrast. *Geology* 19: 1161–4.

Worsley, T. R. and R. D. Nance. 1989. Carbon redox and climate control through Earth history: A speculative reconstruction. *Palaeogeogr Palaeoclimatol Palaeoecol* (Global Planetary Change Section) 75: 259–82.

Wray, G. A., J. S. Levinton, and L. H. Shapiro. 1996. Molecular evidence for deep Precambrian divergences among metazoan phyla. *Science* 274: 568–73.

Wright, V. P. 1990. "Soils." In: D. E. G. Briggs and P. R. Crowther, eds., *Palaeobiology: A Synthesis,* pp. 57–9. Oxford, England: Blackwell Scientific.

Yapp, C. J. and H. Poths. 1992. Ancient atmospheric CO_2 pressures inferred from natural goethites. *Nature* 355: 342–4.

Yapp, C. J. and H. Poths. 1994. Productivity of pre-vascular continental biota inferred from the $Fe(CO_3)OH$ content of goethite. *Nature* 368: 49–51.

Yapp, C. J. and H. Poths. 1996. Carbon isotopes in continental weathering environments and variations in ancient atmospheric CO_2 pressure. *Earth Planet Sci Lett* 137: 71–82.

Yarilova, E. A. 1950. The alteration of the minerals of syenite in the initial stages of soil formation. *Akad Nauk SSSR Trudy Pochvennogo Inst VV Dokuchaeva* 34: 110–42 (translation: Springfield, VA: Clearinghouse for Federal Scientific and Technical Information, U.S. Dept. Commerce, IPST Cat. No. 1360).

Yates, F. E. 1987a. "Physics of Self-organization." In: F. E. Yates, ed., *Self-Organizing Systems: The Emergence of Order,* pp. 409–16. New York: Plenum Press.

Yates, F. E. 1987b. "Quantumstuff and Biostuff." In: F. E. Yates, ed., *Self-Organizing Systems: The Emergence of Order,* pp. 617–44. New York: Plenum Press.

Yatsu, E. 1988. *The Nature of Weathering: An Introduction.* Tokyo: Sozosha.

Young, G. M. 1991. Stratigraphy, sedimentology and tectonic setting of the Huronian Supergroup. Geological Association of Canada, Mineralogical Association of Canada, Society of Economic Geology Annual Meeting, Toronto 91, Field Trip B5 Guidebook.

Young, G. M. 1993. Impacts, tillites, and the breakup of Gondwanaland: A discussion. *J Geol* 101: 675–9.

Young, G. M. amd D. G. F. Long. 1976. Ice-wedge casts from the Huronian Ramsay Lake Formation (>2,300 m. y. old) near Espanola, Ontario, Canada. *Palaeogr Palaeoclim Palaeoecol* 19: 191–200.

Young, G. M., V. von Brunn, D. J. C. Gold, and W. E. L. Minter. 1998. Earth's oldest reported glaciation: physical and chemical evidence from the Archean Mozaan Group of South Africa. *J Geol* 106: 523–38.

Zachar, D. 1982. *Soil Erosion.* Amsterdam: Elsevier Science.

Zahnle, K. J. 1986. Photochemistry of methane and the formation of hydrocyanic acid (HCN) in the Earth's early atmosphere. *J Geophys Res* 91: 2819–34.

Zahnle, K. J. and J. C. G. Walker. 1987. A constant daylength during the Precambrian era? *Precambrian Res* 37: 95–105.

Zhang, C., S. Liu, T. J. Phelps, D. R. Cole, J. Horita, S. M. Fortier, M. Elless, and J. W. Valley. 1997. Physiochemical, mineralogical, and isotopic characterization of magnetite-rich iron oxides formed by thermophilic iron-reducing bacteria. *Geochim Cosmochim Acta* 61: 4621–32.

actinolichen, 135
angiosperms, 96, 137
anhydrite
    as temperature indicator, 109–10
antilichen, 135–6
ants, 44, 46
ammonia greenhouse, 110, 112
Archean
    cyanobacteria, 103–4
    sediments, 108

BLAG, 39–40
bioastronomy, 179–81
biosphere
    end, 182
    evolution, geophysiological model,
        138–42
biotic enhancement, 66–98, 129–31,
    135–8
bryophytes, 124, 136

carbon
    biogeochemical cycle, 16–20
    isotopic record, 26–31
    reaction with seafloor basalt, 24–5
    silicate-carbonate geochemical cycle,
        climatic stabilizer, 21–23, 34–39
    soil, 20

carbon dioxide
    anthropogenic, 17–20
    concentrating mechanism, 113–4
    paleoatmospheric levels
        Phanerozoic, 40–2
        Precambrian, 110–13
    soil, 20, 45, 90–1
chrysophytes, 68, 70
clouds
    Archean, 14, 131
    modeling, 150
continental crust, growth over time, 146,
    152–3

denudation: see weathering
determinism, 3, 125–6, 176
diamond, 117
diatoms, 68, 70
Daisyworld, 8–10
DMS (dimethyl sulfide) and clouds,
    13–14, 131

earthworms, 44
endosymbiogenesis, 3, 141–2, 173–4
entropy
    flow from Earth, 161–70
eucaryotes
    origin, 173–4

faint young sun paradox, 32–3
ferrihydrite, 68–9
frost wedging, 55, 126–7

Gaia
  homeostatic, 5–8, 11, 15, 16,
    172–3
  homeorrhetic, 12
  progressive, 12
GEOCARB, 40, 95–6
geophysiology, 10, 138–42
glaciation
  Carboniferous/Permian, 137
  Huronian, 114–8
  Neoproterozoic, 115
gymnosperms, 137
gypsum
  as temperature indicator, 109–10

habitability
  Earth, 91–5, 181–3
  terrestrial planets, 183–4
habitable zone
  continuously (CHZ), 184–8
hypersea, 176
hyperthermophiles, 103, 132–3, 189

impacts, 114–7

lichen, 48–9, 67–74

Mars, 6, 189
Metazoa
  emergence, 137–8, 174
methane greenhouse, 110, 112
mitochondria, 124, 173–4
molecular phylogeny, 103–6, 132–3

origin of life, 143, 181–2
oxalic acid, 45, 49

oxygen
  atmospheric, 4–5, 119–20, 122
  isotopes
    cherts, carbonates, seawater, 106

paleosols, 4, 54–5, 113
peatlands, 14
pH, ocean, 110
psychrophiles, 118

radiation budget, Earth's surface, 162
Raymo hypothesis, 61–4
regolith, 83, 97–8
rhizosphere, 137

self-organization, 158–60
siderite, 113
soils
  formation, 43–8
  microbial, 53–4
  stabilization, erosion, 51–5
solar luminosity, 32–4
sulfur isotopes, 110, 111
surface area
  BET, 77–8, 91
  geometric, 77–8, 91
  global biota, 45

temperature
  altitudinal variation, 126–7
  diurnal variation, 126–7
  equilibrium, Urey reaction, 158, 171,
    178
  ground, 88–9
  latitudinal variation, 126–7
  upper limit for organismal growth,
    123–4
thermal cracking, 55, 127
thermophiles, 103–5, 113, 124, 132–3,
  189

ultraviolet (uv) screen, 134

Vernadsky, 15–16

weathering
    abiotic, 55, 97–8
    activation energy, 74, 87–8, 127–8
    albedo-enhanced, 89
    alpine, 74
    biogeochemical, 44–6
    biogeophysical, 44
    biomass uptake, 76
    biotic, 55
    biotic enhancement, 66–98, 128–31,
        134–7
    biotic sink, 49–51
    chemical, 44
    fungal, 45, 48
    glacial, 62–3

Iceland, 66–7
intensity
    Archean, Proterozoic, 108–9
laboratory/field paradox, 77–9
lichen, 48–9, 67–74
microbial 53–4
mountain building, 58–61
mycorrihizae, 45, 137
physical, 44
    frost wedging, 55, 127
    thermal cracking, 55, 127
Precambrian, 54–5
reverse, 25–6
rhizosphere, 45, 137
runoff, 88–9
sandbox experiments, 75–6
soil formation, 43–8
WHAK, 34

Printed in the USA
CPSIA information can be obtained
at www.ICGtesting.com
JSHW021438221024
72172JS00005B/48